FOREWORD

In 1969 the Forestry Commission celebrated its Golden Jubilee. It was then felt that the time had come to publish a review of research work carried out by or on behalf of the Commission over its first half century.

From the beginning the Forestry Commissioners recognised the need for research, the necessity for which was especially great in a country in which forests and forestry had been long neglected, and much of the afforestation was to be on bare land in difficult areas. Research therefore began in a small way very soon after the Commission was set up in 1919.

Since that time the research organisation has been gradually expanded (notably in the years just after the Second World War). A large body of work has been done in almost the whole field of forest research and development by the staff of the Commission's own research centres at Alice Holt, Hampshire, and Roslin near Edinburgh, and also, with funds provided by the Commission, at other research stations and in the Universities.

This bulletin has been written by Mr R. F. Wood, who was for 20 years on the staff of the Forest Research Station, Alice Holt, where from 1947 to 1962 he was the senior silviculturist. Then, during the final illness of the late T. R. Peace he served as Acting Chief Research Officer, after which, from 1963 to 1967, he became the Conservator (Research) and deputy to the Director of the Research Division. A photograph of the author appears in Plate 1.

FORESTRY COMMISSION

January 1974

ACKNOWLEDGEMENTS

All the photographs are drawn from the Forestry Commission collection, and have been taken by staff members in the Research Directorate. Collection negative numbers have been added as guides to identification.

Those photos that illustrate the second, post-1945, period are mainly the work of Mr I. A. Anderson, Principal Photographer, and members of his staff. Mr J. Williams, illustrator, designed the cover.

MAPS

Outline maps, showing Forestry Commission Conservancy boundaries, and the locations of all Forests mentioned in the text, are reproduced as Plates 72 to 74.

CONTENTS

LIST OF PLATES

Plate

1. The author, Mr R. F. Wood, with Miss Blanche Benzian.
2. Sowing acorns at Kennington Nursery, Oxford, in 1933.
3. Weed control by blowlamp at Kennington Nursery, Oxford, in 1933.
4. Poplar cultivation. Plants pruned for planting at Kennington Nursery, Oxford, in 1937.
5. Chopping bracken for composting at Kinver Nursery, Staffordshire, in 1949.

6. Nursery weed control with chemicals. Sodium chlorate at Kennington Nursery, Oxford, in 1935.
7. Long-term nursery nutrition experiments at Wareham Forest Nursery, Dorset.
8. Ploughing with horses in *Calluna* ground at Teindland, Laigh of Moray Forest, Morayshire, in 1931.
9. Shallow ploughing with the "Auto-culto" machine on poor *Calluna* ground at Wareham Forest, Dorset.
10. Cultivation experiments at Achnashellach Forest, Ross-shire.

11. Draining experiment at Benmore Forest, Argyll, in 1935.
12. The high elevation experiment at Beddgelert Forest, Caernarvonshire, sixteen years after planting.
13. High elevation at Beddgelert Forest, Caernarvonshire, two years after planting in 1929.
14. Experimental area at the Lon Mor, Inchnacardoch Forest, Inverness-shire.
15. Kilmun Arboretum and Forest Plots from the Dunoon/Benmore Road, seen across Holy Loch, Benmore Forest, Argyll.

16. Method and position of planting experiment at Teindland, Laigh of Moray Forest, Morayshire, planted in 1934.

FORESTRY COMMISSION BULLETIN
No. 50

Fifty [Years of Forestry Research] h

A Revie[w] ted
by th[e]

Ch[...]

LONDON: HER MAJESTY'S STATIONERY OFFICE
1974

1 SBN 011 710142 7

Plate

50. High pruning of Norway spruce at Drummond Hill Forest, Perthshire.

51. Underplanting of larch at Dyfi Forest, Merioneth.
52. Aerodynamic studies at Redesdale Forest, Northumberland.
53. Oak spacing experiment showing 3 ft × 2 ft (1 × 0.66 m approx.) spacing plot, 18 years after planting in 1936.
54. Cuthbertson double-furrow drainage plough at South Kintyre Forest, Argyll.
55. The development of female flowers on Sitka spruce.

56. Measuring moisture content and temperature of soil, using a portable alternating current bridge.
57. Measurement of moisture content of soil using a neutron probe.
58. An elite Scots pine in the Black Wood of Rannoch, Rannoch Forest, Perthshire.
59. Grafting plus tree scion on to sturdy rootstock; waxing has been completed.
60. Larch canker.

61. Stump treatment with fungicide to control *Fomes annosus*.
62. Pine shoot beetle, *Tomicus (Myelophilus) piniperda*, damage to Scots pine shoot.
63. Pine shoot beetle, *Tomicus (Myelophilus) piniperda*, damage to Scots pine at Wareham Forest, Dorset in 1946.
64. Defoliation of Scots pine by Pine looper moth larvae, *Bupalus piniarius*, at Cannock Chase Forest, Staffordshire.
65. Feeding damage by larvae of the Pine looper moth, *Bupalus piniarius*, on needles of young Scots pine shoot.

66. Insect suction trap.
67. Full-grown larva of the Pine looper moth, *Bupalus piniarius*, on a needle of Scots pine.
68. Sitka spruce thinning experiments at Forest of Ae, Dumfriesshire, showing plantations planted at 3 ft × 3 ft (1 × 1 m approx.).
69. Sitka spruce thinning experiments at Forest of Ae, Dumfriesshire, showing plantations planted at 8 ft × 8 ft (2.4 × 2.4 m approx.).
70. Bitterlich's Relascope in use in the forest.

71. The Barr and Stroud Optical Dendrometer.
72. Map: Commission Forests in Northern England.
73. Map: Commission Forests in Southern England and Wales.
74. Map: Commission Forests in Scotland.

INTRODUCTION

This is an account of research work undertaken or supported by the Forestry Commission during the first fifty years of its life.

G. B. Ryle, in his *Forest Service* (Ryle, 1969), gives much interesting information about the establishment of the Commission and its early problems and achievements. Such general matter will only be mentioned briefly here, and only so far as is pertinent to research.

Before the establishment of the Forestry Commission by the Forestry Act of 1919, several Government bodies had an interest in forestry. The Agricultural Departments for England and Wales, Scotland and Ireland handled forestry as a part of the rural economy, encouraged forest education, published some technical matter, and did a certain amount of demonstration and development work. They employed a few professionally trained officers, most of whom at one time or another played a part in forest education. The Office of Woods looked after the old Crown forests, and early in this century began to employ professionally trained staff. The first of these was Roy Lister Robinson; later Secretary to the Acland Committee (on whose report the Act of 1919 was based); Technical Commissioner from 1920 to 1932; and Chairman from then onwards till his death in 1952. W. H. Guillebaud, the leader of research in the Commission from the early nineteen-twenties till just after the second war, also served with the Office of Woods, as did several other "founding members" of the Commission's staff. The immediate technical importance of this continuity was that the Crown Commissioners had advanced from mere maintenance of the old woodlands, and had in fact begun upland afforestation, and both Robinson and Guillebaud had experience of the pioneer efforts at Inverliever Forest in Argyll, and Hafod Fawr in Merioneth (now part of Beddgelert Forest).

In dealing with the research work carried out by or supported by the Commission some mention should be made of what had been done before. The year 1919 marks the beginning of *organised* forest research in Great Britain, and before this there was no concentrated effort and no continuity.

James Macdonald (1954), contributing to the centenary of the Royal Scottish Forestry Society, gave a fascinating account of forestry research in Scotland from 1854 to 1953. He cast his net wide as regards subject, classifying the work under such heads as history, climatic factors (including a surprising intervention by Robert Louis Stevenson!); ecology and soils; forest botany; growth and yield; management; entomology and pathology; utilisation; and silviculture. For the pre-Commission part of the hundred years he was only able to find seventeen published items that he could call forest research, but his paper is admirable for its historic sweep. No attempt will be made to summarise it; James Macdonald hated and would not read summaries! Had he been writing about Britain as a whole he would undoubtedly have included an equal number of items from the other countries. Two obvious ones are the forest gardens planted in England, Ireland and Wales in the late nineteenth and early twentieth century by various authorities as trials of species and forestry demonstrations (Macdonald, 1931), and Augustine Henry's pioneer work at Kew on the hybridisation of forest trees for enhanced vigour (Henry, 1912). But James Macdonald's comment that items of forest research were "isolated incidents" is true for the country as a whole. It is in fact hardly realistic to separate the forest research element from the general legacy of the past; from the plant and soil sciences, arboriculture and pioneer forestry, and from the European professional tradition.

Several of the Commission's publications which deal with research topics contain historical matter, and some of these reviews are quite detailed. The volume already written is considerable, and it is obvious that any comprehensive treatment would be unreadably long. In selecting and summarising, the most useful line is to bring out the main streams, and to give as much attention as possible to contemporary thinking and research approach. But this certainly results in a cavalier dismissal of a great deal of valuable work.

There is also the difficulty of deciding what is research and what is not. The subject matter of this work is *organised research*, but it is not intended to give the impression that "Research" has been responsible for all technical advances. Such is very far from the case. The Commission as a whole has been very much a pioneering body from the outset, and it is only necessary to scan the pre-war numbers of the *Journal of the Forestry Commission* to get a vivid impression of the lively flow of ideas and of the active role of the "field practitioner" in the general advance of knowledge and methods.

Taking the fifty-year period as a whole, it would be difficult to find subjects in which Research has not been engaged at all, but there are several important fields in which most of the progress has been due to field officers. A good example is the methodology of fire protection and suppression. Likewise, pre-war developments in ploughing owed much to the field.

Arrangement of the Contents

It has been convenient to divide the work into two main parts: Part I dealing with research up to the end of the 1939–45 war (for brevity, war in this context is the Second World War); and Part II with the greatly expanded activities from the end of the war till 1970. In each Part matters of organisation and development are treated as far as possible separately from the accounts of the actual research work. Any such division in time is of course somewhat artificial, and will not be adhered to strictly; certain chapters will look forward or back as appears necessary. But the two periods are sufficiently distinct to warrant separate treatment.

Sources and Acknowledgements

A number of the most important sources from the historical point of view are cited in the text, and together with other appropriate references are given in subject bibliographies at the end of each chapter. The Forestry Commission has exercised its right to publish (one of its original terms of reference) from the beginning. Two series were started in 1920; *Bulletins* and *Leaflets*. The Bulletins were designed to be the main carriers for the broader research investigations, and they still fulfil this role. The Leaflets contain technical information on narrower topics, which may or may not emanate from research. Other series have been added since the war, the most important in this context being the published annual *Report on forest research*. Before the war there were no published progress reports on research. The Research Officers rendered typescript reports to the Chief Research Officer, who prepared summarised versions of these, which exist in mimeographed form for most years in the period. From 1930 onwards the progress reports prepared specifically for the Research Advisory Committee provide a very useful series.

The *Journal of the Forestry Commission* (printed for internal circulation 1922–1969) provided the main source of information for field staff on research progress; notes on research and reports on research conferences were a regular feature of the Journal up to the time of the first publication of the *Report on forest research*. Papers on research topics were published in the journals of the Royal Scottish Forestry Society, the Royal Forestry Society of England and Wales, and especially in *Forestry*, the journal of the Society of Foresters, which was founded in 1926, as the "learned society" of the forestry profession. One of the most striking developments in publication since the war is the increase in the number of papers to be found in scientific journals other than those devoted to forestry.

For matters of policy and organisation the Commission's own files have been consulted, and grateful acknowledgement is made to the Forestry Commissioners for this courtesy. Of special value have been all reports and reviews prepared for various purposes by Mr W. H. Guillebaud, the Commission's First Research Officer, who was first appointed in 1919, and retired as Deputy Director General at the Forestry Commission in 1953. His death in 1973, just before this bulletin appears, is recorded with regret. Part I of this publication is very largely based on his writings.

I am very grateful to many erstwhile colleagues for amendments of fact and suggestions for improvements in the text, and am particularly indebted to O. N. Blatchford and S. H. Sharpley for their help in collation and in searching for source material, and to Mrs V. O. C. Lampard for coping efficiently with a much amended typescipt. Mrs R. Mathias prepared the index.

REFERENCES

HENRY, AUGUSTINE (1912). The artificial production of vigorous trees. *J. Dep. Agric. tech. Instruct. Ireland* 15 (1).

MACDONALD, JAMES (1931). *Forest gardens*. Bull. For. Commn, Lond. 12.

MACDONALD, JAMES (1954). Forestry research and experiment in Scotland, 1854–1953. *Scott. For.* **8** (3), 127–141.

RYLE, G. B. (1969). *Forest service: The first 45 years of the Forestry Commission.* Newton Abbot, Devon: David & Charles.

Part I

1920–1945

Chapter 1

POLICY AND ORGANISATION

General Policy

The Forestry Act of 1919, which established the Forestry Commission, empowered the Commissioners to "make, or aid in making, such enquiries, experiments and research, and collect or aid in collecting such information as they may think important for the purpose of promoting forestry, and the teaching of forestry, and to publish or otherwise take steps to make known the results of such enquiries, experiments or research, and to disseminate such information".

The Machinery of Government Committee ("Haldane Report", 1918) had published its remarkably prescient report in 1918. Amongst many other topics, the Haldane Committee considered the arrangements for conducting research of direct concern to Government Departments, and advocated the arrangement often referred to as the "Haldane Principle", under which Government sponsored research is carried out under the general control of the state-financed, but autonomous, bodies known as Research Councils. At this time there was only one such body, the Department of Scientific and Industrial Research (DSIR), established in 1915 (strictly speaking a Department rather than a Council, but acting in most respects as such). The Haldane Report forecast that the growth of scientific research would necessitate the setting up of other Research Councils in appropriate fields, and this has in fact come to pass.

The Commission's early research policy was governed by the Cabinet Committee set up in 1919 to consider the co-ordination of research undertaken by Government Departments. The Agricultural sub-Committee of the main Committee recommended in 1920:

(i)   that a research institution under the Forestry Commission be set up to deal with the problems of the "growing crop";

(ii) that for research on other subjects in connection with forestry problems, and for fundamental research other than that directed to immediate economic benefits, the Commission should first refer to the "appropriate authority" for the particular subject; and

(iii) that a Forest Products Research Board should be set up for research into the utilisation of timber and other forest products.

These recommendations were very much in line with the advice tendered by the First British Empire Forestry Conference (1921), which met in 1920, and this Conference has been influential in matters of forest policy on several other occasions.

Whilst it was not possible at this time, because of shortage of money, to set up an "institute", the Commission proceeded immediately to organise its own research branch; it had in fact taken the first steps in 1919 by the appointments of Research Officers for England and Wales (R. E. Marsden) and Scotland (Dr. H. M. Steven). The second recommendation provided a guiding principle, though this proved difficult to implement in the absence of a Research Council having forest science amongst its specific terms of reference. The third recommendation led to the establishment of the Forest Products Research Laboratory under the aegis of the Department of Industrial and Scientific Research; the laboratory at Princes Risborough, Buckinghamshire, being opened in 1927.

In certain countries (usually those with large forest resources) forest products research is undertaken by the forest authority. The primary role of the new Forest Products Research Laboratory was service to the colonial Empire, and hence it was logical that it should not fall under the Forestry Commission. However, the Laboratory included home produced timbers in its programmes from the outset, and close liaison was maintained with the Commission. After the war, when the production from British forests began to rise rapidly, special arrangements were made with the Laboratory for collaborative research in which the forester's questions (such as how silvicultural practices affect timber quality) received special attention.

Domestic Policy

In Guillebaud's words "the function of the Research Branch was primarily to act as the handmaiden of the executive. It was to carry out experiments designed to improve techniques in all branches of practical forestry work, to study the factors of production, and to hold a watching brief for new diseases and pests as well as discovering methods of controlling existing disorders." Direct service, with all that implies in consultation and laison, was without doubt the essence of the Branch from the beginning. The value of field experimentation is now taken very much for granted, but when the Commission embarked on its programme the approach was new, and the techniques under active development; the reliance of the Commission on experiments was a special feature distinguishing its work from that of many other forest research organisations.

The decision, taken at the outset, that the embryo Research Branch should answer to the Commission (instead of being territorially devolved) was also an important one.

Fundamental Research

This proved a difficult field for the Commission before the second war. Under the 1919 Forestry Act the Commission had authority to support such research financially, where it appeared likely to yield direct benefits to forestry. Funds were never abundant, but as great a difficulty was to find qualified people sufficiently interested, and stationed at institutions suited to the work. Some problems by good fortune could be referred to workers already in the field. For instance, voles (a periodic pest in forestry) were already of interest to Mr C. Elton (later Lord Elton) at the Bureau of Animal Population (Oxford University) in Oxford, and grants were made to him over a period of years by the Forestry Commission. Dr M. C. Rayner at Bedford College had already a name in the study of mycorrhizal relations, and the Commission's grant to her to work on this aspect of the nutrition of forest trees opened a long research story, the later developments of which moved far from the original subject matter. The Commission did not have great success in obtaining DSIR support for fundamental research on forestry subjects.

Between the wars, the University forestry faculties maintained very small staffs, and this restricted the amount of research they could undertake, though valuable work was done, especially at Aberdeen. Oxford, however, where the staff of the Imperial Forestry Institute was integrated with the Department of Forestry, had more scope for research than the other Universities. Early difficulties over the arrangements for fundamental research led to a review of policy by the Research Co-ordination Sub-Committee of the Committee of Civil Research in 1929. Their report recommended the appointment of an Advisory Committee on Forest Research, and the provision of sufficient funds for the Commission to finance fundamental research within their sphere of interest. The Advisory Committee was duly appointed in January 1930, but the recommended increase in funds did not eventuate.

In 1931 the Agricultural Research Council (ARC) was established, and in the division of responsibilities between the new Council and DSIR, it appeared that ARC had become the proper authority for research within the basic sciences common to both agriculture and forestry. This view was followed up; in 1937 the Advisory Committee on Forest Research drew up a list of the many borderline subjects of research, which was submitted to the Agricultural Research Council, who decided to set up a joint sub-committee to study the proposed subject matter, and if necessary, to recommend arrangements for financing it. As it happened, the crises of 1939 intervened and the sub-committee did not meet.

It is historically interesting that basic forest science should have been close to finding itself a Research Council sponsor so many years ago, since it was not till the establishment of the Natural Environment Research Council in 1965 that "long-term forest research" was specifically assigned to a Research Council. The scope and facilities of the various research institutions under the aegis of ARC would have made it technically easy for that body to embrace many subjects basic to forest research, and it may be thought that this solution would have been efficient and economical. As it happened, *direct* assistance by ARC to forest research was restricted to the important subjects of forest soils and tree nutrition, and this was arranged by grants rather than by ARC adoption of the subject matter into its own programmes.

Advisory Committee on Forest Research

This committee, first set up (as noted above) in 1930, has had an important influence on the direction and content of Forestry Commission research over the whole of the period under review. Its membership was:

	From	*To*
Lord Robinson of Kielder Forest and of Adelaide (*Chairman*)	1930	1952
Dr E. J. Butler, C.I.E., D.Sc., M.B., F.R.S.	1930	1935
Sir Arthur Hill, K.C.M.G., F.R.S.	1930	1943
Dr A. F. Joseph, D.Sc., F.I.C.	1930	1932
Sir Guy Marshall, C.M.G., D.Sc., F.R.S.	1930	1945
Sir Ralph Pearson, C.I.E., F.L.S.	1930	1933
Prof. R. S. Troup, C.I.E., F.R.S.	1930	1935
Prof. Sir William Wright-Smith, M.A., F.R.S.E., F.L.S.	1930	1957
G. V. Jacks, M.A., B.Sc.	1932	1963
Prof. J. H. Priestley, D.S.O.	1933	1944
W. A. Robertson, F.R.G.S.	1933	1945
Prof. F. T. Brooks, M.A., F.R.S.	1935	1952
J. N. Oliphant, M.B.E.	1935	1939
Prof. Sir Harry Champion, C.I.E.	1941	1959
Sir Edward Salisbury, C.B.E., D.Sc., F.R.S.	1943	1960
Dr F. Y. Henderson, D.Sc., D.I.C.	1945	1960
Dr S. A. Neave, C.M.G., O.B.E., D.Sc.	1945	—
Dr W. J. Hall, C.M.G., M.C., D.Sc.	1946	1960
Prof. H. M. Steven, C.B.E., Ph.D., F.R.S.E.	1945	1963
Prof. John Walton, D.Sc., F.R.S.E.	1950	1958

	From	To
Sir William Ogg, M.A., B.Sc(Agr.), Ph.D.	1952	1960
Dr J. W. Gregor, C.B.E., Ph.D., D.Sc., F.L.S.	1952	1963
The Earl of Radnor, K.C.V.O. (*Chairman*)	1957	1963
Prof. W. H. Pearsall, F.R.S.	1957	1963
Sir Frederick Bawden, M.A., F.R.S. (*Chairman from 1966 to 1972*)	1960	*Date*
Prof. F. W. Rogers Brambell, F.R.S.	1960	1963
Prof. R. Brown, F.R.S.	1960	1963
Dr D. J. Finney, F.R.S.	1960	1963
Prof. E. C. Mobbs, O.B.E., M.A., B.Sc.	1960	1967
Prof. R. D. Preston, D.Sc., Ph.D., F.R.S.	1960	1963
Dr A. B. Stewart, M.A., B.Sc., Ph.D.	1960	1967
Dr A. C. Copisarow	1961	1963
J. Bryan	1963	1965
The Earl of Waldegrave, T.D., D.L. (*Chairman*)	1964	1965
Prof. M. V. Laurie, C.B.E.	1964	1968
A. R. Wannop	1964	1967
Prof. P. F. Wareing, D.Sc., Ph.D., F.R.S.	1964	*Date*
Dr W. P. K. Findlay, D.Sc.	1967	1971
Prof. W. J. Thomas, M.Sc., M.A.(Econ.)	1967	*Date*
J. F. Levy, B.Sc., A.R.C.S.	1968	*Date*
Prof. J. D. Matthews, B.Sc.	1968	*Date*
Dr N. W. Simmonds, Sc.D., A.I.C.T.A., F.R.S.E., F.I.Biol.	1968	*Date*
Dr R. L. Mitchell, B.Sc., Ph.D., F.R.I.C., F.R.S.E.	1969	*Date*

The first Chairman, Roy Lister Robinson (later Sir Roy; later Lord Robinson of Kielder Forest and of Adelaide), was Technical Commissioner and later Chairman of the Forestry Commission. This was an unusual arrangement, for it put him, so to speak, in the position of advising himself. But in practice it had certain advantages for he could direct the discussions to what he, as the Commission's chief professional forester, knew to be the most urgent problems; also he received the full force of argument. In 1966 the chair passed to an independent member of the Committee.

The composition of the Committee varied naturally in time, but retained much the same pattern. The important connections with the Imperial Forestry Institute, Oxford; Kew; Rotham-sted; the Macaulay Institute and the Forest Products Research Laboratory were strengthened by representation on the Committee, and usually by the heads of these establishments. It was also usual to have a Professor of one of the other faculties of forestry; efforts were always made to have entomologists and mycologists of distinction, and more recently physiology has been represented.

The proceedings of the Research Advisory Committee had a number of regular features. A progress report on research prepared by Guillebaud was presented annually for discussion, together with proposals for new work. It was the custom to discuss wholly new projects with the members more interested. Proposals for new research grants were put to the Committee and advice was often sought about the profitability of continuing particular lines of grant aided work. In the early days, panels of two or three committee members were often asked to vet specialised publications arising out of grant-aided work. One of the most important roles of the Committee was to keep the Commission in touch with scientific developments in allied fields with bearing on forestry problems, and in several cases the initiative towards new approaches in research was provided by the Committee; for instance, the Research Advisory Committee was very much concerned in the early post war developments in nursery nutrition, forest genetics, and forest soils research. An important development in the Committee itself was the setting up of formal sub-committees to study particular lines of work; the first of these, the sub-Committee on Nursery Nutrition chaired by Prof. Brooks, was established in 1944.

The Committee was very much concerned in the research aspects of the Commissioners' Command Paper *Post war forest policy* (Forestry Commission, 1943) which led to the Forestry Act of 1945. The share of the Committee in influencing the organisation of research, or developments in particular fields of research, will be mentioned in Part II under the appropriate topic.

Organisation

For the whole of the period between the wars the Research Branch had a simple structure and a small professional staff. An officer at London headquarters, formally styled Chief Research Officer from 1926 onwards, answered directly to R. L. Robinson (in his role as Technical Commissioner initially, later as Chairman). And except for the period 1923–26, when the headquarters post was held by Prof. A. W. Borthwick, W. H. Guillebaud held this central position. Borthwick covered education and publication in addition to research, but subsequently the two former subjects were taken over by Fraser Story.

Under the Chief Research Officer two Research

Officers covered respectively England-and-Wales, and Scotland, and they were responsible to the CRO for the planning and execution of all nursery and plantation experimental work. One subject specialisation quickly developed, and from 1922 onwards responsibility for the numerous sample plots laid down by the Commission in their own and in private plantations was separated from general research, and given to a "Sample Plots Officer". This separation, which was primarily functional, was necessary in view of the weight of work and the importance of building up statistics quickly on the yield to be expected from forestry in this country. It also stimulated the development of mensurational techniques, thinning studies, and the compilation of forest yield statistics as an aid to management. The separation may, however, have had some drawbacks, in that it tended to restrict silvicultural studies to immature crops.

Till 1928, the Commission retained a consulting Entomologist (Dr J. W. Munro) and a Mycologist (Dr Malcolm Wilson). The system of employing specialists in protection was then dropped in favour of an arrangement with the Imperial Forestry Institute under which the Commission paid grants to the Institute for the services of assistant specialists working to the Institute's senior Mycologist and Entomologist. This arrangement lasted till the establishment of the Commission's Research Station at Alice Holt in 1946.

The principal research appointments for the pre-war period are given below.

It is certain that the mention of individuals of the pre-war team in connection with particular items of work will do very much less than justice to their contributions to research and to British forestry generally. In the relatively small pre-war organisation, which was increasing its knowledge and developing its methods rapidly over a wide field of activity, the leading figures in Research exercised a greater influence than has been the case since the war. Guillebaud's position at the centre of affairs demanded (and received) a breadth of knowledge and an ability to think constructively over a wider range of subject matter than is now required in an age of specialisation. The Research Officers shared this breadth of outlook; none was a narrow experimentalist. Several of course made notable contributions to forestry outside the forest authority; A. W. Borthwick and H. M. Steven in turn held the Chair of Forestry at Aberdeen University, and Mark Anderson that of Edinburgh.

The Research Officers of the pre-war period were all forestry graduates employed in the same "departmental grades" as the field officers; these grades became parallel to those of Professional Classes of the Works Group of Civil Servants. It was not till after the war that certain research specialists were appointed to the Science Group.

For the first few years, Research Officers had no field assistants, but superintended all experimental work directly. In 1922 the first field assistant was appointed, and from the mid-nineteen-twenties onwards a succession of such appointments were made to the principal centres of experiment. These men were nearly all trained in the Commission's own Schools, and were graded on the same lines as the Commission's general corps of foresters; i.e. in the early years of their service they held the old "industrial" ranks of Ganger and Foreman before becoming full Foresters. It was usual for Research Foresters to join Research on leaving their forest school, and most served the Branch for the whole of their careers. Such appointments being relatively few between the wars, Research foresters were picked men, and forest research owes much to this small group, who became masters in the practice of field experimentation.

The contributions of the Research foresters did

Research Staff 1919–1940, including Officers in charge of Sample Plots

Date appointed	Headquarters	England	Scotland	Sample Plots (Whole country)
1919	W. H. Guillebaud (co-ordination)	R. E. Marsden	H. M. Steven (and Sample Plots)	
1920	A. W. Borthwick (Chief Technical Assistant)		M. L. Anderson (Sample Plots only)	
1922		W. H. Guillebaud		M. L. Anderson
1923	A. W. Borthwick (Res., Education & Publications)	H. M. Steven	M. L. Anderson	J. Macdonald
1926	W. H. Guillebaud (Chief Research Officer)		(Direct charge of C.R.O.)	
1928			M. L. Anderson	
1930		M. L. Anderson	J. Macdonald	J. A. B. Macdonald
1932		J. Macdonald	J. A. B. Macdonald	A. M. Mackenzie (then Foreman)
1936		R. G. Sanzen-Baker		
1939				W. R. Thomson (then Foreman)

not stop at the execution, assessment, and maintenance of experiments. They shared to the full in the exchange of ideas, and did valuable liaison and advisory work with their colleagues in the field. This pattern of working with the key position of the Research forester has been one of the most important features of the Research organisation, and has been much admired by other forest research authorities.

Finance

In the period between the wars, the Commission's annual total expenditure on research rose from £6,900 (averaged over first five years) to £18,000 (averaged over last five years). This includes grants in aid of research at outside institutions, which rose from £600 to over £3,000 per annum over the same period. Research expenditure represented something less than 2 per cent of the Commission's total expenditure over this period.

These figures appear very small in the light of current research expenditure (the total figure for 1969 was £873,094), but this should be multiplied by a factor of approximately 4·5 to compare with current money values.

Commenting on the rise in the figure for grants in aid of external research, Guillebaud considered this marked "the passing of the first phase of intensive experiments and the opening of a new phase involving more specialised knowledge and the use of laboratory methods. . . ." In retrospect, he appears to have been previous in this estimate, for the field experimental programme continued to grow after the war in relation to other forms of research.

REFERENCES

BRITISH EMPIRE FORESTRY CONFERENCE (1921). *Proceedings, Resolutions and Summary of Statements.* Forestry Commission, London. (Resolutions published by HMSO as Cmd 865, 1920.)

FORESTRY COMMISSION (1943). *Post-war forest policy. Report by H.M. Forestry Commissioners.* Cmd. 6447. London: HMSO.

"HALDANE REPORT" (1918). *Report of the Machinery of Government Committee.* Cd 9230. HMSO.

Chapter 2

PROBLEMS AND APPROACH

Problems Confronting the Commission

The Act of 1919 established a new national forest authority, but its principal task, afforestation, was new partly because of its scale. Although the Commission's plantations embraced many old woodland areas and other sites which were within existing experience, it was necessary to push out from the normal limits of estate planting onto the poorer upland grazings, heathlands and moorlands, beyond the economic range of farming.

In 1921 the problems confronting the Commission were grouped under four major heads:—(i) establishment of plantations—the term being used in the widest sense to include nursery work, planting methods, species, provenance, etc., (ii) improvement of production—thinning, pruning, etc., (iii) protection, and (iv) the peat soils—considered a separate entity requiring special attention. It was on the first two heads that the Commission's own research staff were expected to concentrate, though the peats were soon taken into the general establishment enquiries.

No detailed scan of the problems confronting the first experimenters will be attempted, but a few general points will be made. Forest nurseries were not new; there has been a nursery trade specialising in forest trees for 200 years at least. But nursery problems grow with the scale of operations and especially with the intensity of working; the need to produce very large quantities of plants quickly, reliably, and cheaply, encouraged numerous lines of enquiry, especially in seedbed technique. Problems concerned with fertility and weed control showed themselves more clearly later on, after long periods of intensive cultivation.

Leaving aside for the moment choice of species and provenance, establishment problems in the forest were of two different orders. Firstly, there were the quite general questions about planting technique, age and type of stock, method of planting, and season of planting etc. Much was known about these subjects, but they had to be re-examined for large-scale operations, always with the aim of keeping the costs of establishment as low as possible. Secondly, there were the cultural questions, not critical for the easy sites, but completely dominating all other enquiries at the difficult end of the range; on heaths and moorlands, and especially on the deeper peats.

Early cultural methods in British forestry have been described by several authors; James Brown's *Forester* (Brown, 1847) is a particularly useful source.

Drainage systems of one sort or another are nearly as old as planting. There are many old references to cultivation (even including agricultural pre-cropping) and ploughing was quite a common practice in the past. But there was little successful experience on the more difficult heathlands and moorlands, and hence the great value of such bold ventures as those of Sir John Stirling Maxwell, at Corrour, Inverness-shire, where Sitka spruce had been planted on elevated peatlands before the first war, and a form of turf planting on the Belgian model had been introduced as early as 1907 (Stirling Maxwell, 1929).

Little or nothing was known about manuring as aid to establishment and early growth, though some trials with slag had started at Corrour before the Commission's time.

All the species which have subsequently been planted in Great Britain on a *forest* scale were in cultivation in 1919, nor can the Commission take the credit for the first plantation of any one of them. However, some (notably Japanese larch and Lodgepole pine) had been but little planted, and the knowledge of the relative performances of species for a wide range of testing environments was most imperfect. Something was known about variation due to origin of seed (provenance), certainly enough to expect it and to admit the problem.

It is of interest that in 1921 "improvement of timber production" seemed most closely associated with such treatments as thinning and pruning. Certainly so far as the *quantitative* aspect of production is concerned, the view has now widened, and it is thought of in terms of all site ameliorations from the establishment phase to the pole stage crop.

The general attitude to forest protection was one of watchfulness, especially towards the new conifers in extensive pure plantations. There was an awareness of troubles on the Continent associated in some schools of thought with monoculture, but the Commission had from the outset a pragmatic approach to such questions.

One important generalisation can be made about all pre-war research work on afforestation: the objective was to find the cheapest effective method of establishing a healthy crop, and not (in modern terminology) to "maximise the profitability of the undertaking". The distinction arises from the broad differences in policy; pre-war policy emphasised the strategic advantages of forestry, whilst post-war policy has placed more weight on the economic aspects. Because of this, pre-war experimentation concentrated on the difficult marginal sites, as a means of bringing large areas of cheap ground into forestry, and not much attention was given to lines of work which might be expected to increase the profitability of easier sites. Perhaps also the atmosphere was unfriendly to "impractical" treatments or levels of such treatments as a means of advancing knowledge, and it is fortunate that the early experimenters went well "beyond the limits" on occasions.

Research Methods

The development of research methods and experimental techniques is a subject in itself, and cannot be dealt with adequately here. Forest research in any particular theme usually starts with observations on what has been done before, and may proceed to ecological studies, formal experimentation or both. Commission research in the early pre-war period was somewhat exceptional in its emphasis on formal experimentation. Of course all pioneering efforts in afforestation were experiments in the primary meaning of the word given in the Oxford Dictionary as "a test, trial, or procedure adopted on the chance of its succeeding", and many early experiments were of this nature. The approach is perhaps apt to be undervalued, but historically at least has played a useful role in raising questions for formal experimentation in the modern sense, that is experiments designed to prove (or more correctly disprove) a specific hypothesis.

The Forestry Commission was fortunate that in the early nineteen-twenties R. A. Fisher and his co-workers at Rothamsted were leading in the development of field plot experimentation and the associated statistical theory. It was to the credit of Robinson, Guillebaud, and especially Steven, that the importance of these developments was appreciated, and that the new experimental methods were applied in forest research at a very early date. Nursery experimentation provided the first (and most suitable) field for these methods developed for agricultural research, and Steven's *Nursery investigations* Bulletin No. 11 (Steven, 1928) is a landmark in the history of forest research methods.

For most nursery experiments, the plot size can be quite small, the square yard being a common unit; hence there are no particular difficulties in applying the principles of randomness and replication. The forest scale is different, and plot sizes (especially for long-term experiments) have to be much larger. Site variation is usually greater than on cultivated land. This may be apparent in the vegetation, or it may be masked by superficial uniformity brought about by fire or grazing. Difficulties in design and layout because of site variability are of course accentuated as the experimental treatments increase in number.

In some early experiments, simple unreplicated comparisons were used where the environment was apparently uniform, and the differences due to treatment were expected to be large, as for instance in species trials on very testing sites.* But otherwise replication was pretty general. Many small plot experiments were laid out with systematic designs, the object being to get "good" comparisons, rather than valid estimates of error. These gave way to more advanced designs with randomisation of treatments; James Macdonald (1933), writing in 1932 on experimental methods in use in Britain, commented that by this time 98 experiments had been laid down in random block designs and 31 in latin squares.

A very high proportion of the experimental work between the wars was concentrated on type forest sites. This was convenient experimentally, and indeed almost essential with limited research staff. It also had the benefit of building up evidence on particular themes step by step, the previous work (and any finer points not easily discernible from the records) being available to the experimenter in a few hours study. These concentrations of experiments also gave some of the advantages of factorial designs (not then available) since treatments might appear in different combinations over the years. The analysis of experimental data was usually restricted to calculating standard errors of means: analysis of variance was not undertaken till after the war.

Many technicalities are involved in the conduct of experiment apart from design; assessments (criteria, sampling methods, units, intervals, etc.); experimental records of all types; labelling systems; etc. etc. In these matters developments owed less to influences outside forestry. Conventions were established relatively quickly. Some were published; Steven's *Nursery investigations* (Steven, 1928) and James Macdonald's *Growth and yield of conifers in Great Britain* (Macdonald, 1928) (Forestry Commission Bulletins 11 and 10 respectively) contain matter on nursery experimental and sample plot procedure. Much however was not published between the wars, but remained in domestic codes. It is worth noting that the "shape" of the forest experimental file (identified by the name of the forest and the number of the experiment in chronological order of establishment) had been established by 1927; it has in fact altered little since then.

Working Arrangements

In a small organisation, such as the Commission's for its first few decades, communication is relatively easy; everyone knows everyone else. But very great care was taken over communication nevertheless. Awareness of problems was very much Robinson's business, and he initiated many of the lines of enquiry

---

* The late Dr E. M. Crowther remarked that one did not need replications to show that bananas would not grow in Yorkshire.

himself. The Chief Research Officer held formal consultations with Divisional Officers every two years or so to discuss experimental work and hear about local problems. Research, or some particular aspect of it, was a common item on the agenda of the Divisional Conferences.

Without field assistants, Research Officers had to superintend all experimental detail themselves. Some very early experiments (such as those on type of stock and season of planting) were carried out by the "Divisions" (the territorial units equivalent to the modern Conservancies), to plans supplied by Research. At this time it is not surprising that the method often failed—in Guillebaud's words "practically all the early experiments carried out by local staff in the Divisions have had to be written off as useless"—foresters were simply too thin on the ground and too busy to give adequate attention to experimental work. The experience lead to the recruitment of Research Foresters, and henceforward Research efforts were concentrated mainly on the "official" experiment, planned, laid out, and assessed by the Research Branch, the co-operation of the field staff usually being confined to the supply of sites and labour.

Practically all the experimental work carried out by the Research Branch between the wars was applied research; "mission oriented" as against "speculative", though perhaps some work on the less common species might fall in the latter category. Where promising results had been attained in experiments, the ideal scheme of progress required that the results should be made known and that they should be tried out on a greater scale or a wider range of conditions. This was usually effected by headquarters circulars drawing attention to the research indications, and calling for field trials of the new method. The results of these gathered together were often printed in the *Journal of the Forestry Commission*, and if of sufficient importance, a change in method might be made a matter of headquarters instruction.

In practice, the progress from experimental result to general adoption did not usually follow this simple path. Simple single-step improvements might be adopted almost at once on the first intimation of experimental results; Guillebaud mentions the introduction of stratification of Douglas fir seed as an example. Relatively few research results can however be translated into straightforward "do this, or don't do that"; most of them raise secondary questions of applicability, some are inter-related with other parts of the system, and some again may change large parts of the system altogether.

In larger bodies, it is often the practice to distinguish the work concerned with the progressing of research results into general practice, and to organise this separately as "development", usually including advisory work with it. Obviously it was quite impossible to take this line between the wars; neither the funds nor the staff were sufficient. But it is clear that in spite of excellent leadership and communications, progress in some lines was delayed for lack of what may be called development work; that is lack of capacity to press on with large-scale and widespread field trials following quickly on the indications of promising experimental results. Guillebaud gives turf planting as an example of the relatively slow adoption of a major change in practice, and mentions fear of initially higher costs and conservatism as delaying factors.

Ryle (1969), referring to the comparatively late arrival of turf planting in South Wales, implies that Steven's early results in Scotland were not pushed hard enough by Research! This is the other end of the same piece of string.

The Commission had in fact a brief experience of one type of development work during 1920–22, when O. J. Sangar held the post of "Co-ordination Officer". Much of what he did would now be called operational research or work study; his most important legacy being an excellent system of forest records. Apart from this, the appointment was probably somewhat previous, and it was not till after the second war that the Commission became seriously interested in development as against research, with the opening of Sections in Machinery Research, Utilisation, and lastly Work Study. And these of course were specialised fields; the more general type of development work concerned with progressing research results into practice remained very much in the hands of the Research Branch.

REFERENCES

BROWN, JAMES (1847). *The forester*. Edinburgh: Blackwood.

MACDONALD, JAMES (1928). *Growth and yield of conifers in Great Britain*. Bull. For. Commn, Lond. 10.

MACDONALD, JAMES (1933). Experimental methods as applied to silvicultural problems in Great Britain. *Proc. Congr. int. Un. Forest Res. Org., Nancy* 1932, pp. 71–89.

RYLE, G. B. (1969). *Forest service: The first 45 years of the Forestry Commission*. Newton Abbot, Devon: David & Charles.

STEVEN, H. M. (1928). *Nursery investigations*. Bull. For. Commn, Lond. 11.

STIRLING MAXWELL, SIR JOHN (1929). *Loch Ossian plantations 1929*. 2 vols. Privately printed (Corrour, Inverness-shire, Scotland, now part of Strath Mashie Forest).

Chapter 3

NURSERY INVESTIGATIONS

At the start, experimental work was carried out at two nurseries near Bagshot (Bushfield and Rapley) in Surrey; at Inverleith, a nursery in the Royal Botanic Garden, Edinburgh; at Seaton and Craibstone, near Aberdeen; and at Beaufort on the Beauly Estate, near Inverness. The well-known Kennington Nursery near Oxford was rented by the Imperial Forestry Institute from St. Johns College in 1927, and made available for Forestry Commission research. Under the supervision of W. G. Gray, it became the chief nursery experimental centre for England and Wales. In Scotland, Inverleith remained an important centre till the early thirties, but work gradually shifted to such nurseries as Altonside, Laigh of Moray Forest, Moray; and the large central nurseries at Newton, Moray and Tulliallan, Devilla Forest, Fife. Hence a difference in method of working evolved, for the Scottish nurseries last mentioned were "production" nurseries, whilst Kennington (and after the war Sugarhill, Wareham Forest) were purely research nurseries. Guillebaud puts the relative advantages very succinctly—"there is something to be said for both methods of working, the independent research nursery has the advantage of better control, and also that a staff can be trained up in experimental methods, but it lacks the contact with the executive side, and the average executive officer or nursery foreman is apt to discount the results on the grounds that the methods are not applicable to large-scale conditions . . .".

Nursery practice lends itself to experiment at all points, and the list of topics on which experiments were actually carried out before the war was very long. For convenience, pre-war research can be considered under the following broad heads:—(i) The seedbed, germination, stocking and survival; (ii) transplant lines; (iii) weed control; (iv) nutrition; (v) pests and diseases; (vi) miscellaneous. Nursery treatments of course interact, and any hard and fast separation is unrealistic.

The Seedbed Plates 2, 3 and 6.
The great preponderance of pre-war experimental work belonged to this general group, the broad objectives being to obtain high and reliable yields of seedlings per pound of seed sown, of suitable dimensions for lining out. A seedling fit for lining out at one year of age was not expected with most conifer species, though it might be achieved with certain hardwoods. This was a general characteristic of the types of nursery in use, most of which (including those in which experimental work was done) having had an agricultural history; they were to be distinguished from the modern "heathland" type of which more will be said later.

To expedite germination various lines were followed. There ·was early interest in actual pre-germination; in fact Squires (1928) practised pre-germination successfully on a large scale in the Windsor nurseries in the nineteen-twenties, and the method was still receiving experimental attention in the early thirties. There has always been an attraction in the idea of removing the germination phase from the hazards of the seedbed, and pre-germination is but the first step to raising stock (for part of the growing period at least) in controlled environments. There has been renewed interest in this recently, but in the pre-war period the line was dropped in favour of seed soaking. A good deal of work was done on stratification, with success in Douglas fir, Lodgepole pine and birch, for which species it became the accepted method of seed pre-treatment by the mid-thirties. No serious work was done in the pre-war period on methods of seed storage, but deterioration in the viability of stored seed had been observed. Routine germination tests were first carried out in the Commission's headquarters in the twenties, but later this service had been passed to the official seed testing stations in Cambridge and Edinburgh, where it remained till the opening of the Commission's own seed testing laboratory in 1951.

There was early recognition of the importance of consolidation and good tilth in the seedbed (normal practice in horticulture and agriculture), but the attainment of really satisfactory physical conditions in the seedbed proved a lengthy business, involving much experimental work on depth and season of sowing, covering materials, shading of seedbeds, etc. In this field of work it is very probable that the modern factorial approach might have speeded things up, since in retrospect the interactions are apparent. Season of sowing experiments for instance (always tricky, and apt to be weather rather than date dependent), did not produce really convincing evidence in favour of early dates (late March, early April) till work on covering materials had dismissed nursery soil, and after drawing blank with such substances as peat, had settled to silt-free sand. This had been pretty well established by 1930 (though first recommended by Marsden as early as 1922), and shortly afterwards attention in Scotland was directed to coarser grits. By 1936 J. A. B. Macdonald had adapted a fertiliser distributor to apply such grits to the seedbed.

Summer shading of seedbeds received a good deal of attention, the experimental results being rather conflicting, but what benefits there were virtually disappeared with early sowing. Much work was also done on the protection of seedbeds from winter frosts, the chief trouble being frost lift, a common and serious pre-war phenomenon. Grit covers and lath shelters reduced it, but essentially frost lift was associated with the small seedling; it ceased to be of significance after the war as the seedling improved.

Transplant Lines

In British forest practice *transplanting* has always been the normal method of producing stock for forest planting, though in certain species (notably oak) seedlings have been favoured. There has often been some interest in the use of seedlings of the common conifers for planting stock, and in processes such as wrenching or undercutting, intended to have similar effects to transplanting on root development.

Experiments carried out at the transplant stage may be concerned purely with the transplanting operation; i.e. using common seedling stock and comparing methods of transplanting (e.g. hand versus "board"), seasons of transplanting, degree of care in transplanting, different spacings in the lines, etc. etc. Otherwise the transplant stage may be a test of seedbed treatments (an "extension" experiment); comparing different sizes, ages or grades of seedling, or seedlings produced by different manurial treatments. These two types of experiment are often combined; and the stock produced by different nursery treatments may be "extended" to the forest as the final test.

A great deal of experimental work on these lines was carried out by the Commission in the nineteen-twenties and described by Steven (1928) in *Nursery investigations*. The survival value of nursery treatments is best mentioned under establishment, and little need be said about nursery experiments as such. The most notable advance in method during the period was the introduction of the lining-out board in the early nineteen-twenties, which is claimed by Alexander (1954) for the brothers Duthie of the firm of Ben Reid of Aberdeen. Much of the work on grading of seedlings had perforce to be repeated after the war with advances in nursery nutrition and the size of seedling. No hard and fast criterion to distinguish a "good" plant emerged, and grading remained a somewhat subjective process. There are of course very great difficulties in separating out the genetic factor in heterogeneous populations from those characteristics which may be ascribed to nursery treatment. But in this connection it should be mentioned that the nursery stages of provenance experiments began to receive organised attention during this period—this was foreseen by Steven as early as 1924 (Lines, 1969)—and one of the earlier provenance experiments on European larch (Anderson's "four-races-four-nurseries experiment") later yielded most interesting information on the longevity of nursery differences as a by-product of the main objective (Edwards, 1954).

Weed Control　Plates 3 and 6.

By present-day standards, pre-war nurseries carried heavy weed populations, and weeds were encouraged by the use of farmyard manure and weed composts. The nurseryman's main defence lay in good husbandry, such as summer fallows and weed smothering green crops; but hand weeding was usually an expensive item and accounted for considerable losses of seedlings from the beds. None of the modern group of herbicides had arrived, and no very great amount of experimental work on weed control was carried out. The blow-lamp (the precursor of the modern flame gun) was, like the lining-out board, an introduction of the brothers Duthie of Ben Reid's in the early nineteen-twenties (Alexander, 1954). As a "pre-emergent" treatment to seedbeds, it gave useful reductions in weeding costs in certain experiments, and entered into common (if not general) practice. It required a good deal of skill and strong nerves. A promising experimental treatment of the early thirties was the use of dilute sulphuric acid watered onto seedbeds as a pre-emergent herbicide, which derived from work at Jealott's Hill on weed control in cereals. This gave a better measure of control than the blow-lamp, and appeared also to stimulate germination, an interesting foreshadow of Crowther's line of work on acidification in the late nineteen-forties. Curiously enough, the well-known complete herbicide sodium chlorate also gave promising results in a few experiments, when applied in extreme dilution as a pre-emergent treatment of seedbeds.

Nutrition　Plates 1, 5 and 7.

Research on nursery nutrition is a long and complex story, and the pre-war part has been attempted already by the present author in his historical notes to Forestry Commission Bulletin No. 37, *Experiments on Nutrition Problems in Forest Nurseries*, written by Blanche Benzian (Wood, 1965; Benzian, 1965). The pre-war period was one of much speculation, but little progress till in the mid and late thirties certain lines of work suggested that a breakthrough was imminent. In retrospect, pre-war experimental work appears to have been dominated by two circumstances. Firstly, there was the widely held belief that the mode of nutrition of coniferous seedlings was special; i.e. they derived their nutriments through the breakdown of organic matter and were ill-adapted to take up the soluble salts contained in agricultural fertilisers; these were considered likely to be damaging. Secondly, there was plain luck; so many of the early probes into direct manuring of the seedbed were made in nurseries which for one reason or another would inhibit any nutritional response in one-year-old seedlings. As it was, no experimental evidence invalidated the former assumption. Perhaps a missed opportunity was the failure to follow up E. V. Laing's work at Aberdeen in the early twenties on the growth of conifer seedlings in nutrient solutions, but his studies were directed mainly towards the peat soils, and the observed fact that conifer seedlings might be grown in dilute solutions of simple salts does not seem to have had much influence on thought about nursery nutrition (Laing, 1932).

There was never doubt about the importance of maintaining the general fertility of nurseries, though evidence in favour of any particular method was hard to obtain. Guillebaud (1937) reviewed the whole question of maintaining fertility with special reference to the use of green or soiling crops, which had been coming into favour especially in Scotland. As it then appeared, the choices lay between a manured green crop phase in the rotation; continuous cropping with organics; and organic manuring with a fallow phase. The second he ruled out on the grounds of poor husbandry (increase of weed populations); it became feasible later with Rayner's weed-free composts. He personally favoured the first alternative.

Dr Margaret Rayner of Bedford College began her researches under Commission grant in 1932. Her work was in fact aimed at the forest rather than the nursery, but the main benefits fell to the nursery. On early heathland planting it had been noticed that both direct sown and natural seedlings of various pines grew quite well close to the stumps of felled Maritime pine. In brief, her assumption was that a normal mycorrhizal association was essential for the proper development of certain conifers, especially on infertile soils, and on some of these (notably the poor Bagshot formations) she suspected the existence of a factor inhibiting normal mycorrhizal association. Thus her work on composts was aimed at cancelling out the inhibiting factor rather than supplying nutrients. Rayner worked with a parallel series of pot experiments at Bedford College and direct forest sowings in small cultivated patches at Wareham, Dorset. The growth of pine seedlings in compost-treated patches at Wareham in 1933 was remarkable; equal success was obtained with pine sowings on hopwaste compost at Allerston, Wykeham Forest, Yorkshire, in 1934. Rayner developed a number of successful composts based on easily obtained bulky organic materials such as straw and bracken, with other organic ingredients to supply sufficient nitrogen for the breakdown process. This was sometimes supplied concentrated as in the form of dried blood, but eventually hopwaste became the key ingredient,

and was either composted pure, or used 25/75 per cent with chopped straw or preferably bracken (Rayner and Neilson-Jones, 1944).

The use of such composts, weedfree, of excellent physical properties, and adequate nutritionally (though the latter was not the main point to Rayner), on acid heathland soils, was the effective origin of the "heathland" or "wasteland" nursery, which took shape with the first cultivation of Sugarhill Nursery, Wareham Forest, in 1943, speedily followed by Harwood Dale, Langdale Forest, Yorkshire.

The interest in organic manuring led James Macdonald to try peat at Kennington in 1934. A half-and-half mixture with nursery soil in board-sided beds and early sowing produced surprisingly large Sitka spruce one-year seedlings (averaging 2½ inch, max. 4¾ inch). Though peat gave usable one-year spruce seedlings in subsequent sowings at Kennington, it failed to do so elsewhere in general trials. The most promising line of advance at this stage appeared to be Rayner's composts, and seed-beds at Kennington prepared from transported wood-land soil heavily composted gave excellent results in 1938. Composts however failed to give results at Widehaugh Nursery, near Hexham in Northumberland, a nursery recently taken in from agricultural land. Rayner's explanation was that agricultural treatments were inimical to the chief mycorrhiza-forming fungi.

There was one indication that coniferous seedlings could respond to soluble fertilisers. Experiments in the late thirties with dilute nutrient solutions containing ammonium sulphate, ammonium phosphate, and potassium nitrate watered onto growing seedbeds gave very good results, especially at Altonside, Laigh of Moray, where the highest level of treatment involved as many as ten applications in the season.

It is of interest that this surge of work directed towards the large seedling in the mid and late nineteen-thirties was closely connected with developments in the field. Turfing and ploughing had increased the immediate costs of establishment; a cheaper plant (perhaps even a seedling) might go some way to balance these extra costs.

During the whole of the pre-war period there seems to have been relatively little speculation on the role of soil reaction in forest nurseries. An intriguing comment by Borthwick (1926) that there appeared to be a "correlation between the intensity of attack by soil fungi and the acidity of the soil" seems to have been an isolated observation. It was made in connection with an experiment at Inverleith, Edinburgh, on the use of wood ash as a seedbed manure, which caused great losses by damping-off.

An important development for Scottish nurseries (and Scottish forest soils research generally) was the arrangement made in 1934 by which the Macaulay Institute for Soil Research, Aberdeen, undertook to provide advisory and other services for the Commission. So far as the nurseries were concerned, this opened a long period during which the Macaulay provided analyses of nursery soils and manurial prescriptions; and for many years Dr A. B. Stewart exercised an important influence over experimental work on nursery nutritional questions in Scotland.

Nursery Pests and Diseases

Pests and diseases in British nurseries have rarely reached catastrophic levels, but have often caused serious losses. The pre-war nursery with its agricultural setting appears to have been ecologically suited to certain insect pests, notably the chafers, and the larvae of two or three species were common pests of seedbeds.

Dr J. N. Munro, the Commission's consulting Entomologist, gave much attention to chafers in the early twenties. Hand picking of the larvae during the cultivating in of green crops was a standard method of dealing with chafers, but experiments were also carried out on the injection of carbon disulphide into seedbeds where damage was seen to be taking place. In practice, this method does not seem to have been generally successful. In the late thirties, J. M. B. Brown (then Assistant Entomologist at Oxford, but since the war the Commission's Ecologist) studied the life history of the various chafers in some detail. With the shift of nurseries to a different ecological setting, chafers have receded in importance, and larvae are seldom seen. Not so much attention seems to have been given before the war to cutworms (larvae of Noctuid moths) which are still with us, though readily controlled with modern insecticides.

The needle cast *Meria* seems to have been troublesome on the larches, encouraged presumably by stocks remaining longer in the nursery than they do now. This is one of the first diseases T. R. Peace studied as assistant mycologist to W. R. Day at Oxford in the early thirties. Sulphur based sprays gave a useful measure of control. Mildew on oak attracted attention periodically, but control measures do not seem to have been considered worthwhile. Losses of seedlings from fungal attacks in the early post-emergent stages (damping-off) were also investigated at Oxford in the thirties. Potassium permanganate, which had been found useful in horticulture, proved disappointing.

Miscellaneous   Plate 4

Besides the work on these broad topics, a number of special projects were pursued. At Kennington for instance a great deal of work was done by W. G. Gray (1953) on the cultivation of the less common species, particularly the broadleaved trees.

Methods of propagation of poplars were pretty well standarised before the war, and this was also a Kennington speciality. Poplars are of course normally propagated vegetatively, but the interests in horticultural circles in the auxins encouraged experimental work in the late thirties on the cutting propagation of less easy subjects, such as elm, plane and even conifers. Even with very primitive propagation facilities results showed some promise, though the auxins did not work miracles. While tree breeding was not in view at this time, Guillebaud had in mind the propagation of disease resistant clones, especially in the elms, since Dutch elm disease had arrived some years before.

One line of work which assumed considerable importance after the war was started in this period; soil sterilisation using formalin was tried at Inverleith, Edinburgh, in 1932, but though there was some slight growth response in seedlings this approach was unfortunately not resumed till 1945.

Nursery experiment assessment is illustrated on the front cover.

REFERENCES

ALEXANDER, J. H. (1954). The forest nursery trade in Scotland, 1854–1953. *Scott. For.* **8** (2), 75–84.

BENZIAN, B. (1965). *Experiments on nutrition problems in forest nurseries.* 2 vols. Bull. For. Commn, Lond. 37.

BORTHWICK, A. W. (1926). Research work. *J. For. Commn* **5**, 10–6 (para. 6).

EDWARDS, M. V. (1954). Scottish studies of the provenance of European larch. *Proc. 11th Congr. int. Un. Forest Res. Org.,* Rome 1953, pp. 432–437.

FORESTRY COMMISSION (1927). *Chafer beetles.* Leafl. For. Commn 17.

FORESTRY COMMISSION (1933). *Leaf cast of larch (Meria laricis* Vuill.) Leafl. For. Commn 21. (Revised by D. H. Phillips 1963).

GRAY, W. G. (1953). *Nursery notes on broadleaved trees.* Res. Brch Pap. For. Commn 10.

GUILLEBAUD, W. H. (1937). Green manuring in forest nurseries. *J. For. Commn* **16**, 20–32.

LAING, E. V. (1932). *Studies on tree roots.* Bull. For. Commn, Lond. 13.

LINES, R. (1969). Early forest research workers in Britain. Letter in *Forestry* **42** (2), 209.

RAYNER, M. C., and NEILSON-JONES, W. (1944). *Problems in tree nutrition.* London: Faber & Faber.

SQUIRES, W. C. (1928). Nursery management and general practice. *J. For. Commn* **7**, 65–69.

STEVEN, H. M. (1928). *Nursery investigations.* Bull. For. Commn, Lond. 11.

WOOD, R. F. (1965). Historical notes. In *Benzian (1965),* 1–5.

Chapter 4

ESTABLISHMENT OF CROPS ON PARTICULAR TYPES OF LAND

Forest experimental work can be classified in at least three ways: (i) the ecological type on which it is done, (ii) the kind of experimental treatment involved, and (iii) the species on which the experiments are performed. Since the chief emphasis in pre-war Commission experimental work was on particular types of land, it will be convenient to discuss the great weight of the work under broad ecological heads, but certain topics which are general to all types will be mentioned in subsequent chapters.

Early reports on research usually distinguish the following broad types: (i) the peats, (ii) the heaths (upland and lowland, including sometimes the East Anglian grass heaths), (iii) the chalk, and (iv) the English clays and loams—considered to be predominantly broadleaved types.

The Peats   Plates 10 to 14

In field experimentation and other research work, the peats undoubtedly account for the greatest effort put into establishment problems before the war. An excellent historical note has already been written on this topic by Zehetmayr (1954). The major figures in the experimental programme before the war were H. M. Steven, M. L. Anderson and J. A. B. Macdonald. In the Commission's early years, the peat soils ranked as a special problem requiring basic research; no *immediate* attack by empirical experimentation appears to have been expected. E. V. Laing and G. K. Fraser both worked on peat problems at Aberdeen University from 1922. Laing's work was largely on root anatomy with special attention to mycorrhizal structures (Laing, 1932), though he also paid some attention to nutrition and physical properties of peats. In retrospect, his observations on root and analyses of plants growing in peat do not seem to have given useful pointers for future experimental work, though undoubtedly adding much to basic knowledge. Fraser was principally concerned at the outset with the physical and chemical properties of peats; this led him to a comprehensive survey and classification of Scottish peat types, published in 1933 as *Studies of Scottish moorlands in relation to tree growth*, Forestry Commission Bulletin No. 15 (Fraser, 1933). This remains the basis for classification of peat covered land types in Scottish forestry.

The key to the evolution of establishment methods on peats was undoubtedly the observation of dependence of root systems on aeration, as evidenced by the death of those parts of the root system buried in the peat, and the extreme shallowness of new adventitious roots, lying just below the vegetation layer. This

was noted by Guillebaud in the early Crown plantations at Hafod Fawr, and specifically described by Steven after studying root systems in one of the 1912 examples of turf planting by Boyd at Inverliever (Steven, 1954). It led to Steven's first experiments on shallow planting in 1924, which included a form of turf planting. Steven did not include phosphate in his experiments at this time, but in 1925 Anderson, who had taken over the Scottish work, followed up the turf experiments, and laid down the first phosphatic manuring trials, which were immediately encouraging.

Experimental work on peats (as on other difficult kinds of planting land) was concentrated on a few type sites, of which in Scotland the Lon Mor area in Inchnacardoch Forest, Inverness-shire, was the most important. Other important areas were Beddgelert in Wales and the Kielder District of Northumberland. Most of the experimental sites were on the difficult fringe, or beyond the margin of plantability in contemporary practice, but variations in the principal areas and outlying experiments in other districts permitted the sampling of a good range of peat conditions.

From the foundation experiments of the mid-twenties research progress was rapid. Turfing and draining (at least superficial draining) are inseparably linked. Sufficient individual turves can be won for planting at 5 ft spacing by draining at about 20 ft intervals and cutting up and spreading the turves. At the other end of the scale the drains (or furrows) may be taken out at planting spacing and the turves inverted alongside in continuous ridges. This of course gives a higher initial intensity of drainage. And turves may be thick or thin (vertical slices laid on their sides). Many combinations were tried, and it was the first mentioned pattern that found its way into general practice on the easier peatlands such as those of the Border region. Anderson, however, foresaw the use of ploughs, and experimented most successfully with shallow turves inverted beside furrows cut at planting spacing (mock ploughing); his most successful pattern broke the line of turves to form a group of three turves at the planting point.

The quickest gains in afforestation practice naturally enough were credited to the easier peats; the *Molinia* and *Juncus* types of fair nutrient content and free from *Calluna*, where the introduction of turf planting revolutionised the establishment of spruces. Concurrently with these improvements in planting technique, Sitka spruce increased its range, and began to assume its dominant position in the moist/elevated/exposed terrain, displacing other species, especially the older exotic Norway spruce. But although turf planting and slag secured a fair take of Sitka spruce almost everywhere, on moorlands with high proportions of *Calluna*, and on the poorer peats typified by *Eriophorum* (*Scirpus*), establishment techniques proved insufficient for sustained growth, and many apparently successful experimental plantations went into "check" before closing canopy. Most of our knowledge about the mechanisms of check in Sitka spruce and about its useful limits in relation to nutrition has been obtained since the war. In certain pre-war experiments, re-application of phosphate where plantations had slowed to near the point of check proved successful. Another line was the introduction of pine nursing. That pines were better starters than spruces wherever heather represented an important component of the vegetation was old knowledge, but formal pine/spruce mixtures aimed specifically at avoiding check in the spruce by the faster start and suppression of *Calluna* by the pines originated, according to Zehetmayr (1954), at Glen Urquhart in 1924. Though such mixtures were tested at several sites by Research, they were hardly a Research introduction. The technique owed much to Robinson's ecological approach; it was he who coined the term "accelerated succession" (in a paper presented to the Research Advisory Committee in 1941) for such mixtures. They proved rather difficult to manage, since in practice there is only a narrow band of conditions in which the spruce benefits; on one side it does not need the nurse, and on the other the nurse grows relatively too fast. Lodgepole pine appears to have made its first mark by being introduced into checked spruce crops, and later it was quite widely used in contemporaneous mixtures. The Commission was somewhat tardy in spotting the usefulness of this species in pure plantation on poor peats and moorlands, and might have learned earlier from the Irish examples. Unfortunately the rough appearance of Washington coastal provenances of Lodgepole pine in mixture with spruce gave the species a bad reputation; had northern coastal seed been available the mixtures would probably have been more manageable.

Phosphatic manures were used before the Commission's time by Sir John Stirling Maxwell at Corrour and also on a small scale by Boyd at Inverliever. The first responses in Commission experiments were observed in trials laid down in 1925. Laing's early work at Aberdeen had suggested that magnesium was necessary for proper mycorrhizal development, hence this element appeared in one of the early trials, but gave only very small responses. Experiments on manuring at time of planting in fact dismissed all the major elements except phosphorus;

basic slag remained in favour for some considerable time, but after the war was largely replaced by ground mineral phosphate, which had done as well as high-grade slag in pre-war trials, and was qualitatively more reliable.

There was an interesting prejudice against soluble forms such as superphosphate. This arose partly from Fraser's work; he said "superphosphate injures the plants, even a year after application" (Fraser, 1933). But there was support for this fear from certain of the Commission experiments using "Semsol"—a mixture of ground mineral and superphosphate—in which considerable losses were sustained, though responses in the surviving plants were good. After the war, it proved very difficult to kill plants with superphosphate, and no doubt the earlier troubles had something to do with the method of application.

None of the pre-war draining experiments was wholly satisfactory in itself, but several gave important indications, and after the war proved invaluable for studies in preparation for the comprehensive programme of draining experiments of the fifties and sixties. Towards the end of the pre-war period ploughing began to make some impact on the peats. (It had appeared a good deal earlier on the heaths.) The first actual experiment in ploughing for draining and turfing was carried out at Borgie, Naver Forest, Sutherland, in 1939, but Maxwell-Macdonald (1935) had taken out plough furrows for turfing using a plough drawn by a tractor-mounted winch in 1934. This proved quite successful on basin peats, but gave trouble on shallow peats over till. He also tried the plough drawn directly by a track-layer tractor, but had not sufficient draw-bar pull. In relatively easy conditions, wheeled tractors such as the John Deere drawing a Ransome Solotrac plough made a fair job of drains for turf planting, as at Twiglees, Dumfriesshire, in 1937/8. The main developments in this field, however, awaited the arrival of wide tracked machines and the special draining ploughs designed by Cutherbertson of Biggar, which equipment belongs to the immediate post-war period.

In addition to the main line peat work, much attention was given to special conditions in the predominantly peat areas. The morainic knolls typical of north-western Scottish scenery provided a special problem. Generally speaking they responded to some form of local cultivation and phosphate, Lodgepole pine being the best species; ploughing eventually overtook this problem. A number of trials were laid down on sites above the normal planting limits. These usually contained comparisons of species, and often mixtures. They proved extremely useful after the war, and many in time became submerged in normal plantations, as establishment techniques improved.

The Heaths Plates 8, 9, 16–20 and 25.

These *Calluna*-dominated lands, drier than the moorlands, with thin peat deposits and podsolised soils, did not present problems of quite the same degree of difficulty as the peat soils. The easier heaths, especially those without very hard pan formations, had been successfully planted with pines (usually Scots pine) in many parts of the country, without any very special establishment technique. Burning-off the heather was a common preparatory step. Short heather might be "screefed" off with the mattock, and this tool might also be used for some local cultivation. However, old accounts (some going back to the late eighteenth/early nineteenth century) mention much more ambitious cultivation, including complete ploughing. No doubt these often referred to transitional heathy conditions. There was also a considerable body of experience in the afforestation of analogous ecological types in northeast Europe, though from the British point of view a great deal of the continental work appeared too expensive, and impossible to copy with the limited finances available.

Large areas of upland *Calluna* heath were acquired in the early twenties, the chief concentrations being in north-east Scotland, and on the north Yorkshire moors. The most important Scottish experimental areas developed in Teindland, Laigh of Moray Forest, Moray, and Clashindarroch, Huntly Forest, Aberdeenshire; in England the greatest concentration of experiments was at Allerston, Pickering District, North Yorkshire. The lowland heaths of the south are much less important in extent; they are scattered and rather variable, being situated on some of the poorest geological formations such as the Bagshot, Lower Greensand, and Culm Measures. (The upland heaths are invariably on till.) The most important of the early experimental areas on lowland heath was at Wareham Forest, Dorset (Hardy's Egdon Heath).

Zehetmayr (1960) has described the experimental work on the upland heaths in detail in *Afforestation of upland heaths*, Forestry Commission Bulletin No. 32. The central theme in the Commission's work on heath was the development of methods of cultivation, in which many field officers played important roles. The first ploughing by the Commission on heath took place in 1921/22 at Allerston, North Yorkshire, using an agricultural tractor and light single furrow plough with subsoiler, which, according to A. H. H. Ross (1931), rarely broke the pan. After a short interval, further ploughing trials were undertaken on the Scottish and Yorkshire heaths and from the mid-twenties onwards progress was continuous.

M. L. Anderson (1951; 1953) originated the "space group" method, and in 1929–32 plantations were established under this system whereby, instead of a constant planting distance, the same number of plants are re-arranged in closely spaced groups leaving unplanted interstices. This "Anderson Group" principle influenced crop composition, spacing and mixtures in post-war years, but in many instances outside trees dominated the groups to confound the intention of producing clean stem final crop trees.

Apart from some early experimental work on hand methods of cultivation, virtually all the experiments involved ploughing, or were sited on ploughed heath. One intriguing non-mechanical approach of the mid-twenties was the use of explosives to fracture the pan locally; this appeared to give some small benefits. Some of the hand cultivation work proved of value in indicating what machines ought to do. The main line, however, was virtually agricultural engineering; the selection of modified agricultural tackle strong enough to perform on heaths without constant breakdowns. The process was by no means complete before the outbreak of war, and the main advances in design of special forestry equipment date from the war years and the immediate post-war period. By the late twenties, however, it was possible to make useful experiments involving such factors as depth of ploughing, complete versus single furrow ploughing, subsoiling, position of planting in relation to ridge and furrow etc. Phosphate had arrived on heaths about the same time as on the peats, i.e. by the mid-twenties; and experimental work on manuring followed much the same course as on the latter types.

Most of the early experiments were carried out with pines since, till ploughing was introduced, these were the only species which could be established. But with ploughing, other species could at least be started, and a considerable number appeared in trials. With ploughing and phosphatic manuring, the heath problem began to appear open-ended; not so much a search for a single solution as on the poorer peats. *Calluna* competition was, however, a much more serious factor for species other than pines, and whatever the ploughing method, *Calluna* tended to return, putting apparently established plantations into check. Exposure was also a selective factor on the more elevated heaths. Sitka spruce was the chief test species in experiments on upland heaths aiming at more productive crops. The first approach in the late twenties was to plant spruce in intimate mixture with pines. The spruce responded very slowly, and it was later appreciated that single plant or row mixtures did not give a sufficient mass of the nurse species to suppress *Calluna* efficiently. An extremely interesting series of experiments of the early thirties were concerned with the use of common broom (*Sarothamnus scoparius*) as a nurse crop, an idea which Guillebaud (1929) had derived from Süchting of Hanover. After dismissing several other legumes, techniques for sowing brooms in bands on plough were quickly worked out. The trees to be nursed were introduced several years later. Such experiments appeared on most of the principal heath experimental areas (including lowland heaths such as Wareham) up to and even after the war. They gave some of the most dramatic results achieved in Commission experiments, but the cost of the technique prevented its wider adoption. An account of this work is given by M. Nimmo and J. Weatherell (1962). The broom nursing effect was undoubtedly complex; heather suppression, shelter, and the leguminous nitrogen fixation mechanism all entering into it. The direct competitive effect of heather was examined in various experiments by removing it mechanically and preventing its return by hand cultivation. This had very marked effects. Killing heather by the use of sodium chlorate was also tried, with partial success. Another interesting line, like leguminous nursing, of continental origin was the application of mulches to suppress the heather round about checked plants. This also had most striking effects, whatever the mulching material used. Much of this work on heather competition was of diagnostic rather than direct practical benefit. It provided a basis for the fundamental studies undertaken at Oxford after the war, and also influenced the Commission's more recent experimental work with herbicides.

Many of the experimental approaches of the upland heaths were also followed at Wareham, Dorset, and Ringwood, Hampshire. These environments were, however, much drier climatically, and Wareham especially proved exceptionally phosphate deficient. The first phosphate responses achieved there seem to have evoked some surprise, since they became apparent unusually quickly. The prospects of species other than the pines never appeared very good on these extreme types, and even Japanese larch, a notably successful pioneer on many upland heaths, proved of little value at Wareham.

Research work at Wareham was prompted by the extremely uneven results achieved by large-scale direct sowings of pine. (Direct sowing is worth some special mention, and will be dealt with later.) Seedlings growing close to the stumps of old felled Maritime pine often showed better than average growth, which might have been explained on physical or nutritional lines, but equally suggested that such seedlings had picked up a mycorrhizal association from the old pine root systems. There certainly appeared to be a case for basic research on pine mycorrhizal associations on these heaths, and in 1932 the first grant was made to Dr. Margaret Rayner of Bedford College for such work. (E. V. Laing at Aberdeen had been working on the mycorrhizal associations of spruce on peat for some time before

this.) Rayner described her researches in several publications, perhaps the most comprehensive account being *Problems in tree nutrition*, written jointly with her husband, Prof. W. Neilson-Jones (Rayner and Neilson-Jones, 1944). As mentioned above under *Nursery investigations*, the main benefits from Rayner's work were in the field of nursery nutrition. Her composts, which were designed to supply nutrients in a form suitable for the formation of symbiotic mycorrhizal associations, were certainly efficaceous in promoting good early growth in the forest. The incorporation of bulky organic manures in patches for direct sowing or planting was of course a most expensive operation, and as concurrently with her work favourable results were being achieved with phosphatic fertilisers, there was never any chance of composts finding any wide application in the forest. In no derogatory sense, a good deal of Rayner's work on mycorrhiza (and E. V. Laing's) proved academic, since under British conditions it has not proved necessary to provide any form of treatment specifically designed to promote symbiotic mycorrhizal associations. This of course is fortunate, and is not the case everywhere.

The arrangements made with the Macaulay Institute for research on Scottish soils had an important influence on later heathland experimental programmes. A. Muir of the Macaulay Institute conducted several detailed pedological surveys of northeastern heathland forests, and G. K. Fraser of Aberdeen (who later joined the Macaulay) collaborated in this work. Their joint publication, *The soils and vegetation of the Bin and Clashindarroch Forests* (Muir and Fraser, 1940) is a most important contribution to the understanding of these environments, and with other surveys enabled heathland experimentation to advance with a much sounder appreciation of the basic soil factors.

The East Anglian "grass heaths", of importance in the pre-war period because of the very large acquisitions of land in Norfolk and Suffolk around Thetford, attracted a limited experimental programme of a special kind. These environments presented no great difficulties for the establishment of pure plantations of Scots and Corsican pines. Taken in from degraded agricultural ground with varying depths of sandy soil over the chalky boulder till (the shallower soils being calcareous, the deeper sands acid and approaching heathland in flora), the vegetation was usually sparse from intensive rabbit grazing. No nutritional problems were met with, and extremely large afforestation programmes were carried out cheaply by planting in shallow plough furrows, which afforded some relief from encroaching vegetation rather than any cultural benefit. On general ecological grounds there was an incentive to diversify the pine plantations. Two series of species trial plots were laid out in the late twenties, one on fairly shallow soils, and the other on the deeper sands. It was found extremely difficult to establish any species susceptible to frost, for this region is noted for the frequency and severity of its late frosts. Pioneer crops of birch and alder established (somewhat painfully) later provided cover for the successful introduction of a considerable range of species, a number of which had failed previously when planted in the open. These species trials were described by R. F. Wood and M. Nimmo (1955). Experiments were also made on the establishment of beech as a potential broadleaved component of the pine stands. In even-aged mixture with Scots pine, beech suffered for years from repeated frosting. Planting into already established pine crops of varying ages indicated that the pine had to be at least six feet high before it afforded much frost protection, and (as elsewhere) introductions into rapidly growing pine crops proved silviculturally impracticable; underplanting in the later pole stage being the first convenient opportunity to introduce frost susceptible shade bearers.

The Chalk

During the late twenties and early thirties several areas of chalk downland were acquired for afforestation. At this time a good deal of the chalk was marginal grazing land, rabbit-infested and neglected; but since the war chalkland agriculture has markedly improved and further afforestation appears most unlikely. From the start, the objective was to establish beech woodland, and most of the experimental work (which started at Friston Forest in Sussex in 1927) was directed to this end. The Chalk downs had a number of features in common with the East Anglian grass heaths. On the removal of the rabbit, dense grass swards returned, highly competitive for moisture, and providing excellent conditions for radiation frosts in the spring and early summer. These factors were of course a much greater handicap to beech than to the pines at Thetford.

Some of the first beech planting on chalk downland was pure, but nursing mixtures became common practice at an early date. European larch was widely used with tolerable success on clay-with-flints, but proved of little value on the thinner soils. Grey alder also did fairly well on the deeper soils. Corsican and Scots pines came into favour later. Ploughing was practised from the start, and field officers tried many appliances and methods. Research was not directly engaged in these mechanical developments, except in studying the effects of cultivation and the obviation of sward competition by hand methods. The generally poor take and slow development of chalk plantations encouraged Research to put the main effort into pioneer crops for beech, and alternative, or subsidiary, species. The survival rates, growth and form

of beech established in pioneer crops of pine were quite clearly superior to those of pure beech or beech grown in even aged mixture, but there were the usual difficulties in manipulating pioneer crops where the beech had been introduced at too early a stage. One of the less common species tried at Queen Elizabeth Forest, Hampshire, *Alnus cordata*, proved exceptionally fast growing. Numerous ancillary experiments connected with establishment were carried out also on such topics as age and type of plant, season of planting etc.; and a few manurial experiments were done with negative results. A survey of chalk plantations conducted by W. R. Day and R. G. Sanzen-Baker in 1938 added much to the understanding of growth of beech in relation to the soil and other environmental factors. Day also made observations on the nutritional disorder lime-induced chlorosis, which was of common occurrence on rendzinas and old cultivations with much free lime in the profile. It eventually caused considerable losses in Scots pine, but not usually before this tree had played a useful pioneering role.

The Commission's general and research experiences were described in *Chalk downland afforestation* by R. F. Wood and M. Nimmo (1962).

The English Clays

The inheritance of the old Crown woodlands, such as the Forest of Dean, Alice Holt, Salcey in Wymersley Forest and Dymock in Hereford Forest, and the acquisition of a number of predominantly oak woodlands felled in the first war, presented the Commission a complex set of problems. Many of these areas were on heavy clay soils, either the geological clays such as the Weald, Gault or Oxford clays, or on clay soils derived from till. Ecologically, the woodlands were oak or oak/ash communities, but often represented the remains of the old management system of coppice with standards, hazel being the chief coppice species, though in the west oak itself had often been managed on coppice rotations. The Commission's early policy was to grow oak where the evidence was that oak timber could be grown to suitable dimensions and quality, but other broadleaved species such as ash and sycamore were also favoured where conditions were considered suitable. Poplars were also tried on heavy soils in old woodland areas, with minimal success. Certain broadleaved experiments were also carried out on lighter soils.

The central theme in the pre-war broadleaved experimental work was the establishment of the new crop; not, as after the second war (when further devastations had presented a similar set of conditions), the treatment of the coppice and other woody regrowth. Oak itself accounted for the greatest share of the experiments, and a more unrewarding subject is difficult to imagine. Seedlings did as well as transplants, but no cultural measures at planting or after would make the oak grow any faster. Ash on the other hand responded briskly to post-planting treatments such as hoeing and nitrogenous dressings. Many experimental direct sowings of acorns were carried out, comparing patch with line sowings, densities and season of sowing etc. These were often very successful, but mice represented an incalculable hazard. Accepting the slow growth of oak plantations as inevitable, various lines were followed with economy as the main object. Spacing experiments in regular plantation had hardly given their results before the pure oak plantation was outmoded, but certainly indicated that the customary very close spacings, such as 4 ft by 2 ft, were quite unnecessary. Much more interesting and productive were Guillebaud's studies on arrangement, where he endeavoured to find the minimum number of plants required in small groups to give at least one well-formed stem—the spacing of such groups also being compared. Larch was used as a nurse with such group plantings and also in band mixtures with oak. These mixtures were locally successful (there are particularly good examples at Crumbland, Monmouthshire), substantial volumes of larch up to timber dimensions being obtained whilst achieving an adequate distribution of well-formed oak stems. Some nursing effect in enhanced height growth of the oak was also demonstrated, but it is a pity that experimental attention was not given to oak/Norway spruce mixtures, since this combination gave good results in practice.

Towards the end of the pre-war period a few trials were made of group planting of various species in broadleaved coppice, a precursor of the post-war programme on Derelict Woodlands. There were many sidelines in the broadleaved programmes; the effort to grow walnut may be mentioned, which at least resulted in positive results on the nursery front, but the tree proved extremely difficult to establish in woodland conditions, and the work may be said to have confirmed its orchard status.

REFERENCES

ANDERSON, M. L. (1951). Spaced group planting and irregularity of stand structure. *Emp. For. Rev.* **30,** 328–341.

ANDERSON, M. L. (1953). Spaced group planting. *Unasylva* **7** (2), 55–63.

FRASER, G. K. (1933). *Studies of Scottish moorlands in relation to tree growth.* Bull. For. Commn, Lond. 15.

GUILLEBAUD, W. H. (1929). Green manuring in forestry. *J. For. Commn* **8,** 66–68.

GUILLEBAUD, W. H. (1930). Hardwood plantations. *J. For. Commn* **9,** 14–22.

LAING, E. V. (1932). *Studies on tree roots.* Bull. For. Commn, Lond. 13.

MAXWELL-MACDONALD, J. (1935). Mechanical drainage. *J. For. Commn* **14,** 65–66.

MUIR, A., and FRASER, G. K. (1940). The soils and vegetation of the Bin and Clashindarroch forests. *Trans. R. Soc. Edinb.* **60,** 233–341.

NIMMO, M., and WEATHERELL, J. (1962). Experiences with leguminous nurses in forestry. *Rep. Forest Res., Lond.* 1961, 126–147.

RAYNER, M. C., and NIELSON-JONES, W. (1944). *Problems in tree nutrition.* London: Faber & Faber.

ROSS, A. H. H. (1931). Moor ploughing. *J. For. Commn* **10,** 33–37.

STEVEN, H. M. (1954). Root form on peat. In *Zehetmayr* (*1954*), 95–109.

WOOD, R. F., and NIMMO, M. (1955). Trials of species in Thetford Chase Forest. *Rep. Forest Res., Lond.* 1954, 106–114.

WOOD, R. F., and NIMMO, M. (1962). *Chalk downland afforestation.* Bull. For. Commn, Lond. 34.

ZEHETMAYR, J. W. L. (1954). *Experiments in tree planting on peat.* Bull. For. Commn, Lond. 22.

ZEHETMAYR, J. W. L. (1960). *Afforestation of upland heaths.* Bull. For. Commn, Lond. 32.

Chapter 5

GENERAL SILVICULTURAL ENQUIRIES

Special types of land attracted the greatest share of experimental work on the establishment of plantations. Some topics, however, were common to several types of land.

Planting Stock

Experiments on the age and type of planting stock and on the grading of stock of common nursery treatment were carried out for a number of species in the pre-war period. Other experiments on planting stock compared plants of the same age and type raised in different nurseries, or by different treatments in the same nursery. And besides the "official" experiments on these topics, extremely large-scale trials were conducted by the field staff on such subjects as grading of transplants, and the use of seedlings versus transplants. The common theme in all this work was to find the cheapest stock, survival rates in the forest being the main criterion, but early growth rate also being a consideration, especially on weedy or frosty sites, where the size of plant might in itself be important. Work on planting stock was often linked with cultivation, method of planting, and season of planting.

From the experimenter's viewpoint, this class of work is often very unrewarding, and the results difficult to interpret. In the British climate and with most of the species used, naked rooted stock will usually survive given reasonable care in planting. It has often proved extremely difficult to demonstrate "intrinsic" differences in the capability to survive between very dissimilar planting stocks under experimental conditions, whereas in general practice one of the kinds of stock may fail altogether. This is particularly the case in comparisons of seedlings and transplants. Many experiments and some large-scale trials indicated that seedling stock of the common conifers was a distinct possibility for forest planting, but on a number of occasions extensive plantings of seedlings proved disastrous. To generalise, pre-war work certainly suggested that the normally higher root/shoot ratio of transplants had its importance, and equally that a certain minimum mass was an insurance factor, but it did not arrive at any well-defined standards of quality. Nursery treatments (other than age/type differences) were not usually critical for survival, mainly because they did not have much influence on the physiological condition of the plant, unlike for instance the modern practice of cold storage. An interesting exception which proved the rule was the dramatic improvement in survival rates in the difficult planting subject *Cupressus macrocarpa* brought about by pre-lifting and shallow heeling in of the plants.

Season of Planting

This is another subject on which it proved difficult to obtain clear cut experimental evidence. Over the period October to April, the actual season of planting is less critical than the state of the plant, the soil conditions, and the weather immediately after planting. Season of planting experiments were usually carried out by planting at more or less fixed dates (e.g. mid-month) over the period to be investigated. They gave (when sufficiently repeated) rough actuarial information about the seasons, rather than clear indications of the best conditions, and that for large-scale afforestation is usually good enough, since the planter may not have great freedom of choice in the matter. The question becomes of greater importance when the species concerned is difficult, or when the conditions are extreme. An interesting illustration of the distinction between seasonal and planting conditions was provided by an English series of experiments on Corsican pine, the major species which has been subject at all times to the greatest planting losses. Averaged over a number of seasons, the safest period appeared to be November to February, but far the best results were obtained in October, with, however, the occasional disaster in seasons when soils had not re-wetted by the planting date. Autumn planting in fact gave favourable results for many species, and but for inflexibility in nursery stock-taking and plant allocation procedures, would undoubtedly have been adopted more widely. It is probably not the best season in very exposed situations. Cold-storage facilities which enable planting stock to be held dormant have greatly prolonged the safe planting season.

Direct Sowing

Robinson was much interested in this establishment technique, and initiated large-scale trials in England and Wales in 1922, Maritime and Corsican pine being the chief species sown. The early methods were somewhat crude, seed being sown after heather burning and some rudimentary cultivation, but there was sufficient local success to retain interest in the method, and it was adopted as a Research project in the late twenties. Although trials of direct sowing appeared on many types of land, including the peats and even sand dunes, the greatest promise was shown on the lowland heaths, and in the pre-war period most of the research work was conducted at Wareham, Dorset.

Direct sowing shared in the general cultural advances associated with heathland research. With the early complete shallow ploughing at Wareham, a compact sowing surface was obtained by running a tracklayer tractor at right angles to the ploughing direction. Early experimental sowings were in drills. As the ploughing method evolved to deep single furrow, direct sowing changed from drills to patches —little platforms cut out from the sides of the plough ridge.

Bonemeal was the first phosphatic manure used. It was later compared with basic slag (though not on a strict phosphorus content basis) and could not claim any great advantage to offset its higher cost. With the commencement of Rayner's investigations, some of her more successful composts were used to manure direct sowing patches in comparison with slag and bonemeal. As the composts were worked in at rates of 2 lb per square foot, there was of course an extra measure of cultivation. Early growth with composts was much better than with the phosphatic fertilisers, but these initial advantages had usually disappeared inside ten years. There was, in any case, no prospect of using compost on the forest scale. A number of pines were raised very satisfactorily at Wareham by direct sowing, and in the pre-war period the costs of the method were not so high as to rule it out. For certain types of lowland heath the question of direct sowing as a method of establishment remained open till well after the war. Experiences with the technique at Wareham were described by R. F. Wood and M. Nimmo (1952).

Spacing in Plantation

It is somewhat surprising that this topic attracted so little attention during the early pre-war period. A few experiments were laid down in the early twenties, but the main experimental attack on the subject was delayed till 1935, when James Macdonald drew up a plan for a series of experiments. This was a good example of what was known as the "Divisional" experiment, in which the plan was drawn up by Research and the execution entrusted to the field officers. Over 150 separate experiments comprising unreplicated plots comparing four or five initial spacings, in many species, were laid down. Unfortunately there was very great wastage in these experiments due to wartime neglect. The best of them, however, were taken over after the war by Research.

Tending

Some experimental work was conducted in the pre-war period on weeding of plantations and pruning for timber quality. Weeding is an awkward subject to deal with experimentally; it is of course easy enough to compare the presence and absence of weeds, but degrees of weeding are very difficult to define. One example of the extreme treatment was hoeing in young ash, and another the removal of *Calluna* in heathland establishment work; at the time both of these might be regarded as diagnostic rather than practical treatments. Degree of weeding was chiefly important in young hardwood plantations, where the establishment period was likely to be prolonged, particularly with oak. The usual experimental approach included intervals between weedings and the extent to which competing vegetation was cut back, e.g. generally, or locally only. Variability of the competing vegetation was the chief difficulty, since an answer for one type might be inappropriate for another not many yards away with different components. A further complication was that weed growth might be beneficial at one season and harmful at another; for instance experienced foresters have claimed that bracken can provide useful shelter at certain times of the year. In broadleaved forestry, the greatest progress was made late in the period when the use of cover began to be appreciated—this became systematised in the early post-war period by O. J. Sangar and R. H. Smith. Later of course the herbicides came on the scene.

The traditional practice of brashing young conifer plantations for ease of access and as a precaution against fire was accepted from the outset, and apart from some practical investigations into tools, no research was done on it. Over the whole of the pre-war period the assumption was that pruning for timber quality was desirable, and would ultimately be justified by increased prices for clear timber and by removing some of the prejudice against knotty home-grown plantation timber. Various enquiries were made into the efficacy and costs of different methods and tools. Certain early experiments compared numbers of stems pruned, and as these involved pruning stems of different degrees of dominance, they later provided valuable indications of the number of trees which could be pruned without undue wastage amongst those losing dominance before putting on useful amounts of clear wood. The pruning of green branches was regarded with some suspicion—especially in the spruces—since it had been associated on the Continent with stem cankers. Apart from a case at Drummond Hill, Perthshire, there was little British evidence that green pruning was a harmful practice. (Dallimore, who as an arboriculturist had pruned specimen trees of many conifer species, always contended that the danger was much over-rated.) The other important aspect of green pruning—the direct effect on growth rates due to the reduction of live crown—was studied with more precision after the war.

Pruned stems from experiments done in 1931 were sent to Princes Risborough in 1944, and useful

Plate 1. The author, Mr R. F. Wood, with Miss Blanche Benzian, who worked on nursery nutrition problems and is the author of Forestry Commission Bulletin No. 37, *Experiments on nutrition problems in forest nurseries*.

X151.

Plate 2. Sowing acorns at Kennington Nursery, Oxford, in 1933. The great preponderance of pre-war experimental work in nurseries was devoted to seedbed studies. Ch. 3.   D790.

Plate 3. Weed control by blowlamp at Kennington Nursery, Oxford, in 1933. As a pre-emergence treatment to seedbeds this tool gave useful reductions in weeding costs and entered into common, if not general, practice before the war. It required a good deal of skill and strong nerves. Ch. 3.   D803.

Plate 4. Poplar cultivation. Plants pruned for planting in the field being lifted at Kennington Nursery, Oxford in 1937. Propagation of poplar under W. G. Gray was a Kennington speciality. Ch. 3. D870.

Plate 5. Chopping bracken for composting at Kinver Nursery, Staffordshire, in 1949. Work on organic manures was carried out by Dr M. Rayner of Bedford College, who developed a number of successful composts based on easily obtained bulky organic materials such as straw and bracken. The use of such composts was the effective origin of the "heathland" or "wasteland" nursery. Ch. 3. B718.

Plate 6. Nursery weed control with chemicals at Kennington Nursery, Oxford, in 1935. The right-hand bed has been treated with 1% sodium chlorate. Applied in extreme dilution as a pre-emergence treatment of seedbeds, this complete herbicide gave promising results in a few experiments, at a time when the modern group of herbicides had not appeared. Ch. 3. D840.

Plate 7. Long-term nursery nutrition experiments at Wareham Forest Nursery, Dorset. Plots were treated year after year with the same organic or chemical fertilisers, with polythene sheeting buried around the boundaries of each plot. It required some twelve seasons of continuous seedling cropping before the performance of seed beds manured purely with fertilisers began to fall regularly below those receiving organic manure. The tall, thin labels in front centre of each plot indicate applications of organic manure (dark label) or chemical fertiliser (light label). Ch. 13. B1588.

Plate 8. Ploughing with horses in *Calluna* ground at Teindland, Laigh of Moray Forest, Morayshire, in 1931. The main line of interest in cultivating *Calluna* heathland before the war was the selection of modified agricultural tackle, strong enough to perform without constant breakdowns. Ch. 4. D1543.

Plate 9. Shallow ploughing with the "Auto-culto" machine on poor *Calluna* ground at Wareham Forest, Dorset. The furrow turned was too shallow to be of much value, and this type of ground preparation gave poor results even with Scots pine and Corsican pine, though it was superior to planting directly into the natural surface. Ch. 4. D39.

Plate 10. Cultivation experiments at Achnashellach Forest, Ross-shire. Turfing and draining are inseparably linked and many combinations were tried during the 1920s. The four plots shown above are, from left to right: (1) Deep drains at 20 ft. (2) Shallow drains at 20 ft. (3) Lazy (fallow) beds above mattock trenches. (4) Deep drains at 9 ft. Ch. 4. D3603.

Plate 11. Draining experiment at Benmore Forest, Argyll, in 1935, showing layout of draining plots with unplanted ground between and turves spread from drains which are at various spacings. Ch. 5. B1718.

Plate 12. The high elevation experiment at Beddgelert Forest, Caernarvonshire. The famous "latin square" planted between 1250 ft and 1800 ft, sixteen years after planting in 1929. Ch. 4. C229.

Plate 13. High elevation experiment at Beddgelert Forest, Caernarvonshire, two years after planting in 1929. Plots run from 1250 ft to 1800 ft, laid out as a classical "latin square". Although too early to show tree growth the area stands out clearly from the surrounding mountainside, following the prevention of sheep grazing by fencing and the 2-year grass growth. Ch. 4. D2.

Plate 14. Experimental area at the Lon Mor, Inchnacardoch Forest, Inverness-shire. A Scots pine provenance trial is shown from centre foreground to left. Alternate lines of Sitka spruce and Lodgepole pine are planted on Belgian and shallow turfs in the right foreground. This is one of the areas where work on peat was concentrated. Ch. 4. D600.

Plate 15. Kilmun Arboretum and Forest Plots from the Dunoon/Benmore Road, across Holy Loch. Benmore Forest, Argyll. The first plot was planted in 1930 by the staff of West Scotland Conservancy, who were responsible for all the work until 1948 when the Research Division took over its management. Ch. 6. A4020.

Plate 16. Method and position of planting experiment at Teindland, Laigh of Moray Forest, Morayshire, planted in 1934. Lodgepole pine, Scots pine, Japanese larch and Sitka spruce set into four different soil horizons: original surface, leached layer, pan, and subsoil. 1934. Ch. 5. C2160.

Plate 17. Oregon alder planted on complete ploughing, with basic slag on the left-hand side and no manure on the right. Experiment at Wykeham Forest, Yorkshire, in 1934. Pioneer crops of alder later provided cover for the successful introduction of a considerable range of species, a number of which had previously failed when planted in the open. Ch. 4. 425.

Plate 18. Phosphate manuring. Applications of basic slag on western hemlock at Teindland, Laigh of Moray Forest, Morayshire. Basic slag applied at the time of planting to the right-hand plot 30 years previously at 10 cwt per acre, but not in immediate foreground. The use of basic slag was much favoured before the war, but was largely replaced by ground mineral phosphate in the post-war years. Ch. 4. C3437.

Plate 19. Nurse crops of common broom (*Sarothamnus scoparius*) to provide nitrogen nutrition on heathland sites. These experiments on most of the principal heath experimental areas, up to and even after the war, gave some of the most dramatic results achieved, but the cost of the technique prevented its wider adoption. The experiment shown above was at Broxa, Langdale Forest, Yorkshire, and shows Sitka spruce and broom sowings on complete "RLR" ploughing which received basic slag at planting and sowing. Ch. 4. D2906.

Plate 20. Heather pulling experiment at Loch Ossian Plantation, Corrour, Strath Mashie Forest, Inverness-shire. On the left, heather pulled with no slag. On the right, heather pulled plus basic slag. Phosphatic manures were used before the Commission's time by Sir John Stirling Maxwell at Corrour. Ch. 4. B2317.

Plate 21. Oak spacing experiment, showing $6\frac{1}{2}$ ft $\times 6\frac{1}{2}$ ft (2×2 m) spacing plot 18 years after planting at Fleet Forest, Kirkcudbrightshire. Ch. 7. C2339.

Plate 23. Provenance trial of Lodgepole pine at Millbuie, Black Isle Forest, Ross-shire, in 1939. Hazelton, British Columbia, provenance. Ch. 6. B2677.

Plate 22. Provenance trial of Lodgepole pine at Milbuie, Black Isle Forest, Ross-shire, planted in 1938. The form of the whole stem and crown are comparatively fine and growth slows up after 22 years. The provenance here is Long Beach, Washington. Ch. 6. D50113.

Plate 25. Mixture of Lodgepole pine and Sitka spruce on *Calluna* at Wykeham Forest, Yorkshire, 11 years after planting in 1922. The formal pine/spruce mixtures aimed specifically at avoiding check in the spruce by the faster start and the suppression of *Calluna* by the pines. They proved rather difficult to manage as the nurse frequently grew relatively too fast. Unfortunately the rough appearance of Washington Coastal provenance of Lodgepole pine in mixture with spruce gave the species a bad reputation. Ch. 4. B1693.

Plate 24. Provenance trial of Lodgepole pine at Millbuie, Black Isle Forest, Ross-shire. An extensive plot of Terrace, British Columbia provenance, planted in 1939. Ch. 6. B2680.

Plate 26. Lodgepole pine provenance trial at Kirroughtree Forest, Kirkcudbrightshire, planted in 1939. In view of its exceptional range no pre-war experiments on Lodgepole pine could be considered adequate, but they were sufficient to indicate the great distinction between coastal material and provenances of more continental origin. The plots above are: Left foreground, Salmon Arm (with many failures). Left background, Queen Charlotte Islands (regular, but poor in height). Right foreground, Smithers (regular, of moderate height). Right background, South-west Lincoln County (regular, fast growth). Ch. 6. C2368.

information on such factors as rate of diameter growth, size of "snag" and condition of branch (dry or green) on speed of occlusion was obtained. This was followed up by a major study of pruning occlusion by C. P. Scott immediately after the war. The economics of pruning (effect on grading etc.) remained to be studied much later, when suitable material from the early experiments became available.

REFERENCES

ANDERSON, J. W. (1932). Summer planting of spruce. *J. For. Commn* **11**, 53–55.

FORESTRY COMMISSION (1936). *Experiments on:* I. *Season of planting Corsican pine.* II. *Season of direct sowing of acorns.* Res. Circ., For. Commn 3.

GUILLEBAUD, W. H. (1933), Pruning in plantations. *Q. Jl. For.* **27**, 122–150.

PINCHIN, R. D. (1951). Nursery extension experiments at Radnor Forest. *Rep. Forest Res., Lond.* 1950, 24–27.

SIMPSON, A. (1933). Corsican pine planting. *J. For. Commn* **12**, 38–39.

WOOD, R. F., and NIMMO, M. (1952). *Direct sowing experiments at Wareham Forest, Dorset, 1928–1949.* Res. Brch Pap. For. Commn 5.

Chapter 6

SPECIES AND PROVENANCE

In any afforestation programme the choice of species is regarded as the most important of all management decisions. The traditional experimental approach to the selection of exotics has three stages: (i) the arboretum, in which tolerance of climate is the main criterion, (ii) "forest gardens" or "forest plots", in which subjects showing promise in the arboretum are studied as forest crops, and further information obtained on ease of establishment, rates of growth, and health; and (iii) trials of species proper, in which precise comparisons are made between a few strong candidates for some particular environment. (Plates 15, 22–25, 28, 29.)

Provenance studies should be an integral part of any programme on exotics, but in Britain and elsewhere provenance work has often been undertaken only after the main species investigations have been well advanced. Whilst pre-war work on species provided examples of each of the principal stages in the enquiry, the most important exotics certainly did not find their places as a result of any orderly series of trials. The European species, larch, Norway spruce and the native Scots pine had long been in cultivation as forest trees, and the nineteenth-century introductions from other regions were well represented in arboreta; indeed virtually all of the species which were to prove important to the Commission had found their way into forest plantations, though some were only to be seen in small trial plots.

A good deal was known about the requirements of the old plantation species, and the use of vegetation types as a guide to the choice of species was a familiar concept. The problem was to fit in the newer exotics to the best advantage inside and outside of the traditional planting range. It was much more an evolutionary process than organised research, and a great deal was learned about the useful limits of species simply by planting them, and especially by over-extending them. Research played a part in this *solvitur ambulando* process by special studies and observations; all the early Research officers were engaged in this in addition to their experimental work. Mark Anderson in particular devoted much study to the British wasteland types, and his *Selection of tree species* (Anderson, 1950) is a most valuable ecological approach to the choice of species for particular sites.

Apart from this general approach to the species question, Research was engaged in numerous more direct enquiries. It was not necessary to go back to the arboretum stage for any great number of species; hence, purely from the forestry point of view, large comprehensive arboreta were not required. The Commission, however, had a general educational interest in forest botany, and it was appropriate enough that they should collaborate with Kew in the establishment of the National Pinetum at Bedgebury Kent, in 1924, which will always be associated with the name of William Dallimore. The forest plots stage was of greater importance, since many trees, coniferous and broadleaved, had previously appeared only as single specimens in arboreta, and had not been grown as forest crops. Two very large forest gardens were established: Bedgebury Forest Plots in 1929 adjacent to the National Pinetum, and Kilmun, Argyll, in 1930. Bedgebury was managed by Research from the outset, but Kilmun was managed by the West Scotland Division (Conservancy) till 1948, when it became a Research project. The objects in the two series were much the same— to represent under forest conditions as many species as possible. As the climates of the two localities are (for Britain) markedly different, the two series provide most interesting contrasts, the most striking being the success of eucalypts at Kilmun, few of which would survive one winter at Bedgebury. One similar private venture should be mentioned, the Forest Garden at Crarae, between Inveraray and Lochgilphead in Argyll, planted by Sir George Campbell, and very generously presented by to him the Commission in 1956. (Plate 15 shows Kilmun.)

Complementing these large near-comprehensive forest gardens, and representing much the same stage in the study of exotics, were numerous smaller collections of species in special environments. Unreplicated series of half a dozen or more species appeared on most of the important types in the pre-war period, plot sizes varying from a tenth of an acre to a whole acre. The third stage was usually represented by the incorporation of two or more obvious candidate species in replicated experiments, comparing different establishment treatments, but a number of replicated "straight" species trials were also laid down.

In retrospect, pre-war Research work on the selection of species had little to do with the identification of the chief afforestation species, which more or less emerged from pre-Commission and early Commission experiences, but it did greatly assist in evaluating them and determining their range. What it did do most successfully was to demonstrate the wealth of species capable of forming productive crops on all but the most restrictive sites, and this was not merely of academic interest in this period, for contemporary thinking favoured diversification of the forests.

It was appreciated that success in afforestation of bare land depended very largely on the use of species with the characteristic of appearing early in the natural succession in their native habitats, but there does seem to have been some confusion of thought (which has not entirely disappeared) in relating the demands of a species to its order in the succession, successor species being usually considered more demanding than pioneers.

It had proved possible in Britain to make successful use of exotics with relatively little attention to seed source, mainly because the readily available seed sources, or those selected for new large-scale introductions, happened to provide material with wide enough tolerances in relation to British conditions. There was no doubt an element of luck in some cases, but with such important species as Sitka spruce and Douglas fir good choices were made with the help of the Canadian authorities. In any case provenance investigations were contemplated at an early date, and Lines (1970) has pointed out that Steven drew up a model plan for provenance studies in 1924 for Douglas fir and Sitka spruce. An extremely important administrative step had been taken in 1922 in the introduction of the system of labelling each seed lot by the year of collection and an individual identity number; these numbers together with the origin of the seed and (later) the germinative characteristics being printed in the Commission's *Journal* (and later reprinted as Research Branch Paper No. 29. *Seed identification numbers*). Although identity was often lost in the passage of plants through the nursery to compartments in the forest, or confused by admixture in beating-up operations, the system frequently provided the first clear indications of troubles related to seed source.

The Commission's first provenance experiments were laid down in the late nineteen-twenties and early thirties. Much the best of the early experiments were those concerned with Scots pine and European larch; the representation of provenances in the pre-war Douglas fir and Sitka spruce experiments being rather meagre, though the Radnor 1929 unreplicated experiment comparing four provenances of Sitka spruce ranging from the Queen Charlotte Isles to California later provided one of the striking demonstrations of clinal variation in growth rates. Naturally, the great weight of the evidence from pre-war provenance work was gathered in the post-war period, but in both Scots pine and Sitka spruce there were early indications of the importance of keeping within a few degrees of our own latitudinal range, since there were considerable difficulties in establishing plants of some of the more extreme provenances.

An interesting early observation was that the progeny of the old Breckland hedgerow Scots pines grew well not only on its home ground at Thetford, but also under very different conditions in Scotland. Later, a good deal of attention was paid to identifying flexible material of this sort in other species.

Of the other important exotics, fairly comprehensive experiments for Norway spruce were laid down in the mid and late nineteen-thirties, though several experiments were lost by fire at Kielder shortly after the war. In view of its exceptional range, no pre-war experiments on *Pinus contorta* could be considered adequate, but they were sufficient to indicate the great distinction between coastal material and provenances of more continental origin. Incidentally, this appears to have been noted, and acted upon, by the Irish a good deal earlier than in Britain, on evidence from very simple contrasts in seed origin at the Avondale Forest Garden, Co. Wicklow, Eire.

Edwards and Howell (1962), in outlining a programme for trials of exotics, have made the point that a new introduction should never be judged on a single seed source, and that a reasonably safe preliminary indication may be obtained from three provenances: (i) from the part of the range where the species reaches its optimum, (ii) from the best climatic match in the natural range, and (iii) (since identity is hardly to be found) from a third source which represents a logical approach to the climate of the planting region. It is surprising how much is covered by this simple and ingenious rule.

REFERENCES

ANDERSON, M. L. (1950). *The selection of tree species.* Edinburgh: Oliver & Boyd.

BROWN, J. M. B. (1954). Profile characters and natural vegetation as guides to the choice of species for wasteland afforestation. *Proc. 11th Congr. int. Un. Forest Res. Org.*, Rome 1953, 297–323.

EDWARDS, M. V., and HOWELL, R. S. (1962). Planning an experimental programme for species trials. Paper for *8th Br. Commonw. For. Conf.*, *East Africa*, For. Commn, Lond.

LINES, R. (1967). *The planning and conduct of provenance experiments.* Res. Dev. Pap. For. Commn, Lond. 45.

LINES, R. (1970). The beginnings of provenance studies. *J. For. Commn* 36 (1968–69), p. 81.

MACDONALD, JAMES (1931). *Forest gardens.* Bull. For. Commn, Lond. 12.

STEVEN, H. M. (1940). Choice of tree species in the north-east of Scotland on the basis of soil and vegetation types. *Forestry* 14, 81–85.

Chapter 7

MENSURATION

During the pre-war period (and in fact up to 1960) Mensuration was a Research Branch activity. Except during the first two or three years when the Research Officers' duties included mensuration, the subject was regarded as a specialisation, and placed under the charge of a Sample Plots Officer for the whole country. A few permanent sample plots had been established by the Crown Office of Woods before the first war, but the Commission, in beginning its own studies of yield and growth, had a more valuable body of statistics from the surveys conducted by the Board of Agriculture and the Board of Trade during the latter part of the first war. During the course of these surveys over 1,000 plots in various parts of the country were measured, and it is of interest that the methods employed in the surveys were published in 1919 as the first of the Forestry Commission's series of Bulletins. In 1920 Bulletin No. 3 was published based on these data and entitled *Rate of growth of conifers in the British Isles*, providing the first reasonably comprehensive set of yield tables for the principal conifer species grown in Britain at that period. Bulletin No. 3 was the work of W. H. Guillebaud (1920), who had in fact been engaged on the original surveys. Apart from these British data, the Commission had of course the published yield tables of various European authorities. No great amount of information was available, however, from any source on the yield of the newer species in plantation, though some useful indications were available from individual plantation records.

The Commission's primary objective in this field of work was to build up a body of information on the growth and yield of the principal species throughout the country: such information to be made available in improved yield tables. The yield tables in Bulletin 3 were prepared from temporary plot measurements, and gave no reliable information on thinning yields, hence one of the first tasks was to establish sufficient permanent sample plots to study the development of stands under standard treatments over the length of the rotation.

An equally important object was to study the effects of different thinning treatments on the development of stands, stem dimensions, form and quality, and on intermediate and total yield. It was also expected that permanent sample plots would provide matter for soil research, and eventually timber of known history for forest products research. Mixed as well as pure stands were included in the programme. (Plate 21)

In forest mensuration, the Commission had the benefit of continental theory and practice. James Macdonald, who held the post of Sample Plots Officer from 1924 till 1930, did most of the early constructive work on British sample plot methods, and in his *Growth and yield of conifers in Great Britain*, F.C. Bulletin 10 (Macdonald, 1928). The practices which were to be followed during the pre-war period were described, with their derivations and reasons for modifications to suit British conditions. One of the key decisions was the choice of a suitable criterion for ranking stands by productivity into quality classes. Without sufficient crop records this could not be done on production itself—the logical method advocated by Baur—but production appeared to be closely enough related to height growth for stands to be grouped into classes according to their height/age curves. The method proved convenient and consistent enough for most purposes, though certain aberrations were noted before the war which warranted further investigations in the post-war period. Cajander's method of classifying sites by fertility as indicated by the vegetation type had little chance of proving useful in Britain in the almost complete absence of a climax vegetation.

A notable contribution to the study of stem form of conifers was made by James Macdonald (1932–34) in the early thirties. Another important matter was the classification of the individual trees in the stand, which had a bearing on the sampling procedure for measurement, and also on the definition of thinning grade. A system modified slightly from that of the Swede Schotte was adopted, which divided stems into classes primarily according to their degree of dominance in the canopy. Thinning methods and grades had long been the subject of controversy on the Continent, but the actual definition of grades had reached a measure of international agreement.

The primary distinction between thinning practices which had become generally recognised in the early years of the century was between low thinnings and crown thinnings. Broadly speaking low thinnings concentrated on the successive removal of the dominated classes, whilst crown thinnings were made amongst the dominants, leaving the dominated classes to form a lower canopy. Each principal type of thinning was sub-divided according to weight, i.e. the quantity of stems removed, and these grades were defined strictly by the classes of stem which could be removed in a particular thinning. By 1923 the Commission had standardised its own experimental thinning grades (based very largely on international conventions), recognising three grades of low thinning and two of crown thinning. A few modifications were made in later years, but pre-war experimental thinning was substantially on the above lines; the control being primarily qualitative and only secondly quantitative.

Historically speaking, this attitude to thinning owed much to continental experience in handling heavily stocked (often naturally regenerated) crops, in which the differentiation of canopy classes might be very marked, and the choice of treatment between low and crown thinnings a very obvious one. The long-term objects of the two main types of thinning were also different; crown thinning aiming at irregularity of stand structure, sometimes with the intention of preparing the way for some form of continuous forestry, whilst low thinning aimed to favour the most vigorous components of the crop, to produce a regular stand structure, and hence even assortments of produce. Of the major conifer species in British forestry, only one (Norway spruce) can be considered a true tolerant, and hence it is not surprising that (as James Macdonald remarked) the development of an under canopy is not well marked in British woodlands. In all the more light-demanding species, pre-war thinning experiments indicated that the standard crown thinnings would not maintain a healthy lower canopy. In the extreme light demanders, such as larch, underplanting at a suitable stage with a tolerant species was a well established practice, and a number of sample plots were established to study the productivity of such two-storeyed crops. The best known of these were the underplanted larch crops at Dymock (Hereford Forest, on the Gloucestershire-Herefordshire border), which have been described by D. E. Earl (1951). Even-aged mixtures were scarcely touched in the pre-war sample plot programme in the absence of suitable comparisons; it remained for M. V. Laurie to apply a strict experimental approach to this subject after the war.

In general the pre-war sample plot programme could not take the vexed subject of stand structure and productivity very far, indeed it still remains a topic for controversy, and some aspects of it call for sophisticated physiological approach. But on the simpler question of relative *weights* of thinning, a good deal of evidence indicated that productivity was not greatly influenced by grade of thinning, and there was more than a suggestion that heavier thinning (within the range studied) increased production by a small amount. Such effects could not emerge as statistically significant till the arrival of the replicated thinning experiment. The first of these, a Latin square of four treatments, was established in a 20-year-old Norway spruce crop at Bowmont Forest, a privately-owned property near Kelso in Roxburghshire, in 1930, and has been justly regarded as an international landmark in this type of enquiry.

Considering the resources available, the pre-war achievement in this field of work was remarkable. Nearly 270 permanent plots were established in twenty species, providing a wealth of information for the post-war studies on growth and yield. Many of the earlier plots were established on private estates (a most valuable contribution by the owners), and some of these plots have been particularly useful in sampling the higher production range on sites not often acquired by the Commission.

REFERENCES

EARL, D. E. (1951). *A short note on the underplanted European larch plots at Dymock, Gloucestershire, and Haldon, Devon.* Unpub. MS, Forestry Commission, Planning and Economics Branch.

GUILLEBAUD, W. H. (1920). *Rate of growth of conifers in the British Isles.* Bull. For. Commn, Lond. 3.

MACDONALD, JAMES (1928). *Growth and yield of conifers in Great Britain.* Bull. For. Commn, Lond. 10.

MACDONALD, JAMES (1932–1934). The form of the stem in coniferous trees. *Forestry* **6**, 53–62 and 143–153; **7**, 32–42 and 121–129; **8**, 56–63.

Chapter 8

FOREST PROTECTION

The term embraces all damaging agencies: biological (insects, fungi, mammals, etc.); climatic (frost, wind, etc.); and last but not least, fire. During the pre-war period organised research concentrated mainly on insects and fungi, but late in the period studies on frost and on soil conditions predisposing to root disease appeared in the programmes. It is curious that no organised research on fire was attempted, since certain aspects of fire protection and suppression would certainly have been open to experimental approach. However, methods evolved fairly quickly by trial and error and exchange of ideas, as did many other techniques not subject to formal enquiries. The relatively low flammability of Japanese larch, and its power of suppression of ground vegetation, was one important field observation which passed into practical use. A comment by J. M. Murray (1931) which seems to have been well ahead of its time was that fire rides should be "burned when the surrounding dead vegetation is too damp to burn readily". He considers this could be done by the use of flame guns with pack sprays in support, but also envisaged the use of fire retardant materials on the margins. The arrival of the desiccant herbicides and "wet" and "viscous" water encouraged experiments in this direction thirty years later.

The only considerable piece of work on mammals during the period was Elton's grant-aided study on the vole (Elton *et al.*, 1935), which threw a great deal of light on the creature's habits and on its extremely variable population. A few efforts were made to find suitable seed dressings to deter mice and birds, and the standard seed dressing, red lead, was a rather curious legacy of this work, for it was not very effective in its intended role, but proved convenient as a colouring agent to facilitate seed distribution. It is only very recently that it has been superseded.

Forest entomology, and (to a slightly less degree) forest mycology, were however active subjects of research from the outset. It has been mentioned previously that up to 1928 the Commission retained its own consultants in entomology and mycology (Dr J. W. Munro of Kew and Dr Malcolm Wilson of Edinburgh University respectively), but later arrangements were made with the Imperial Forestry Institute for the coverage of these subjects.

According to James Macdonald (1954) forest entomology had been actively studied (especially in Scotland) for a considerable period before the Commission came into being, and many of the more important insect pests had been described and appreciated. In forest pathology, A. W. Borthwick (who became the Commission's first Chief Research Officer, and later held the chair of forestry at Aberdeen) had made some of the first scientific studies of tree diseases in Britain during the early years of the century. One specific disease, canker of European larch, had collected an extensive literature from early times.

No insect pests during the pre-war period threatened to reach disaster level, but several were the cause of very significant economic losses. The extensive fellings of the first war provided ample breeding material for the weevil *Hylobius* and the bark beetles *Hylastes* and *Tomicus*, and considerable damage was caused to new plantations on old conifer sites. Munro worked on log trapping methods for *Hylobius*, which gave a fair measure of protection. He made a comprehensive study of the bark beetles, and published *British bark beetles* as Forestry Commission Bulletin No. 8 (Munro, 1926). With relatively little felling of older conifers in the late twenties and thirties, and with adequate forest sanitation, these pests receded in importance for a time. There were occasional alarms about the introduction of non-indigenous bark beetles in unbarked pitwood, but no bark beetle in this period appeared likely to be other than a secondary pest. (Plate 27.)

Britain has been fortunate that the principal European defoliators, though present, have rarely proved of importance. (The Pine looper, *Bupalus piniarius*, proved the exception in the fifties (Hussey, 1957)). The Oak leaf roller moth, *Tortrix viridana*, frequently defoliated oak in the early summer, but without lasting damage (Hussey, 1961). Much more alarming were effects on the young East Anglian pine plantations of the related tortricid *Evetria buoliana*, the Pine shoot moth, which was ranked as a serious hazard in the early thirties (Forestry Commission, 1956). The hollowing out of young shoots resulted in much deformation of crowns, including the classic symptom of "post-horn" stems, once very common, but now rather rare. This moth attracted a great deal of study, and one of the remedial measures tried was the hand picking of inhabited buds, which did not prove effective. Eventually, it appeared that the damage (though obvious enough) did not reduce the choice of stems to a serious extent, and the high populations of the moth declined as the wave of large-scale plantings in the area passed through the early thicket stage.

The complex groups of the sap-sucking Aphids and Adelgids were studied over the whole of the period, these being a speciality of R. Neil Chrystal who wrote the Commission's Bulletins Nos. 4 and 7 on the *Douglas fir Chermes* and the *Silver fir Chermes* respectively (Chrystal, 1922; 1926). The latter, now known as *Adelges nordmannianae*, provide our sole example of an insect pest precluding the use of an otherwise useful exotic (*Abies alba*, the European Silver fir). Adelgid attack on European larch was often very conspicuous, and was regarded as playing some part in the complex die-back syndrome. *Adelges cooleyi* on Douglas fir also caused very disfiguring attacks, but it was noted that these did not usually account for more than slowing of growth for a few years.

The Green spruce aphis, *Elatobium abietina*, does not seem to have been a matter of very great concern over much of the pre-war period, possibly because there was not then so great an acreage of Sitka spruce at a vulnerable stage, and sited in over dry environments.

In the early part of the pre-war period the targets for research in mycology (or more generally, forest pathology) do not seem to have been so clear as those for the entomologist. Though several diseases had yet to make an appearance in Britain, all the principal diseases were familiar enough to mycologists and had long been described. What was not clear was their probable status and relative importance in extensive afforestation with much reliance on exotics. Diseases specific to the more important "new" species were a subject for special attention. For instance the increasing incidence of canker and die-back due to the fungus *Phomopsis* on Douglas fir (and to a less extent on Japanese larch) caused a good deal of concern in the nineteen-twenties, and Malcolm Wilson made a special study of this trouble (Wilson, 1925). This particular disease did not however turn out to be so damaging as had been feared, and to this generation it may seem odd that it attracted the first Commission Bulletin on a pathological subject.

Malcolm Wilson had seen Dutch elm disease on the Continent, and expected that it would find its way to Britain. It duly arrived, the first well-proven case being at Totteridge, Hertfordshire, in 1927. Curiously enough, this outbreak was confused by deaths resulting from a leaking gas main. Dutch elm disease was the first subject on which T. R. Peace worked for the Forestry Commission when he was appointed to the Imperial Forestry Institute as assistant pathologist to W. R. Day under the new arrangement. He kept the disease under observation for many years, and published *The status and development of elm disease in Britain* as Forestry Commission Bulletin No. 33 (Peace, 1960). He was not himself able to find very strong evidence for specific or varietal resistance, but some indications of this amongst British elm populations, coupled with the promise of the Dutch selections, provided one of the special arguments for opening genetical work on forest trees. Dutch elm disease presented the first

well attested example of an insect-carried fungal disease in British forestry.

Peace also worked before the war on heart rot of oak in the Dean Forest, caused by *Stereum* species, and showed that the trouble was associated with the prevalence of large dead limbs; a common symptom of heavy early thinnings and subsequent neglect. He also started his work on poplars and varietal resistance to canker, which developed into a major project after the war.

W. R. Day, who was Pathologist in the Department of Forestry at Oxford from the creation of the post in 1924 till his retirement in 1961, worked over a wide range of subject matter before the war. His approach was primarily ecological, with special interests in soil conditions and climatic factors predisposing to disease. With Peace he did the first British work on testing of provenances (of European larch) for resistance to freezing temperatures in the laboratory. This particular study was not conclusive, and Day's opinion that frost damage played a major part in Larch canker was based mainly on his detailed field observations. The notorious May frost of 1935 provided a great opportunity for Day and Peace to study the incidence of frost damage on many tree species in relation to stage of growth, topography, and other environmental features, such as cover. *Spring frosts*, Forestry Commission Bulletin No. 18 (Day and Peace, 1937), was the first important British text on forest meteorology. (Plate 60.)

Day conducted several detailed surveys in connection with establishment difficulties or poor growth. His work with R. G. Sanzen-Baker on the Chalk has already been mentioned, and he also collaborated with Sanzen-Baker in an investigation into poor growth in the South Wales plantations of Llanover, Ebbw Forest, and Llantrisant, Tair Onen Forest, where the difficulties had been ascribed to atmospheric pollution. Their finding was, however, that most of the troubles were due to incorrect choice of species and poor establishment practices.

No very great amount of work was done before the war on the root diseases. Day investigated Ink disease of Sweet chestnut, due to a *Phytophthora*, and found anaerobic soil conditions a predisposing factor. He was inclined to the view that even *Armillaria mellea* might be secondary, though he himself had found the fungus penetrating apparently healthy roots. Relatively little mention was made of *Fomes* in the pre-war period, as the destructive potential of this fungus only became clear after the first thinnings of plantations on susceptible sites, such as the old agricultural soils at Thetford.

REFERENCES

CHRYSTAL, R. N. (1922). *The Douglas fir Chermes, Chermes cooleyi*. Bull. For. Commn, Lond. 4.

CHRYSTAL, R. N. (1926). *The Silver fir Chermes.* Bull. For. Commn, Lond. 7.

CHRYSTAL, R. N. (1936). *Studies on the Pine shoot moth, Evetria buoliana* Schiff. Bull. For. Commn, Lond. 16.

DAY, W. R., and PEACE, T. R. (1937). *Spring frosts, with special reference to the frosts of May.* Bull. For. Commn, Lond. 18.

ELTON, C., DAVIS, D. H. S., and FINDLAY, G. M. (1935). An epidemic among voles (*Microtus agrestis*) on the Scottish Borders in the spring of 1934. *J. Anim. Ecol.* **4**, 277–288.

FORESTRY COMMISSION (1956). *The pine shoot moth.* Leafl. For. Commn 40.

HUSSEY, N. W. (1957). Effects of the physical environment on the development of the Pine looper, *Bupalus piniarius. Rep. Forest Res., Lond.,* 111–128.

HUSSEY, N. W. (1961). *Oak leaf roller moth.* Leafl. For. Commn 10.

MACDONALD, JAMES (1954). Forestry research and experiments in Scotland, 1854–1953. *Scott. For.* **8**, 127–141.

MUNRO, J. W. (1920). *Survey of forest insect conditions in the British Isles, 1919.* Bull. For. Commn, Lond. 2.

MUNRO, J. W. (1926). *British bark beetles.* Bull. For. Commn, Lond. 8.

MURRAY, J. M. (1931). Clearing of fire rides. *J. For. Commn* **10**, 5–6.

PEACE, T. R. (1960). *The status and development of elm disease in Britain.* Bull. For. Commn, Lond. 33.

WILSON, M. (1925). *The Phomopsis disease of conifers.* Bull. For. Commn, Lond. 6.

Chapter 9

WARTIME EMERGENCY ARRANGEMENTS

Under the appropriate chapter heading *The great divide*, Ryle (1969) describes the separation of the staff of the Commission into two groups, one concerned with timber production (and eventually forming the core of the Timber Production Department of the Ministry of Supply), the other continuing the Commission's normal operations as best they could. Research continued, but the work of the Branch was very much curtailed. W. H. Guillebaud took over the administrative charge of the Forest of Dean, but retained his duties as Chief Research Officer. Likewise J. A. B. Macdonald doubled the role of Divisional Officer for the West of Scotland with that of Research Officer for Scotland. R. G. Sanzen-Baker continued as Research Officer, England, but also did a great deal of organisation of volunteers from the Universities. Several of the experienced Research foresters were temporarily transferred to forest management duties for the war period. Hence the capacity to plan and execute new work was strictly limited, and the shortage of experienced forest workers made it difficult to maintain forest experiments adequately. Some forest experiments undoubtedly suffered, but generally speaking the great body of experimental work came through the war period very well, much to the credit of the short-handed corps of Research foresters. Probably the most serious wartime losses were due to forest fires in the dry spring of 1942; many experiments were lost at the important Scottish sites at Teindland (Laigh of Moray Forest), Inchnacardoch and Borgie, and the forest plots at Bedgebury, Kent, also suffered severely.

No wholly new experimental projects were started during the war, but Guillebaud's group planting in degraded coppice in Monmouthshire marked a change of direction in the treatment of such areas which was followed up after the war in the Derelict Woodlands project. Nursery investigations continued actively, and by the end of the war ideas on nursery nutrition had been revolutionised by the success of Rayner's sowings on compost.

As in the first world war, the extensive clear fellings of woodlands all over the country offered opportunities for the collection of information. But whereas in 1914–18 the surveys of fellings provided most useful information of growth and yield, which was badly wanted at that time, the Commission's own network of permanent sample plots had by now largely taken care of that particular object. However, Prof. Champion of Oxford University took the initiative in organising a survey carried out by the teaching staffs of the Universities with the objects of: (i) correlating rates of growth with locality factors; (ii) studying to incidence of heart-rot; and (iii) estimating the effects of stem defect on the out-turn of produce. A great deal of data was collected, but various difficulties prevented the whole being worked up and published. Probably the most valuable practical result of the survey was the information obtained on the incidence and extent of damage caused by *Fomes*, which influenced Peace in placing high priority on *Fomes* in the post-war pathological programmes.

Another by-product of wartime fellings was a study of the air seasoning of stacked pitprops assessed by periodic weighments, which was organised by Guillebaud. His figures for weight loss in the dry summer of 1940 indicated rather quicker rates of seasoning than in previous investigations by the Forest Products Research Laboratory.

Of greater ultimate importance than any forest research undertaken during the war were the various decisions about post-war research policy and the major lines of research. The war was a particularly active period for the Commission in planning forest policy; the Command Paper *Post war forest policy* was presented to Parliament in 1943 (Forestry Commission, 1943), preparing the way for the Forestry Act of 1945.

The Research Advisory Committee continued to meet throughout the war, and the ideas behind the re-shaping of forest Research for the post-war period are fully reflected in its proceedings. It will be convenient to refer to some of the arguments when dealing with the topic concerned in Part II.

REFERENCES

FORESTRY COMMISSION (1943). *Post-war forest policy. Report by H.M. Forestry Commissioners.* Cmd 6447. London: HMSO.

RYLE, G. B. (1969). *Forest service: The first 45 years of the Forestry Commission.* Newton Abbot, Devon: David & Charles.

Part II

1946–1970

Chapter 10

POST-WAR POLICY

In considering the general policy for research immediately after the war in the Research Advisory Committee, references were made to the special role of the Agricultural Research Council, and to the proposed discussions with that body and DSIR on the responsibility for fundamental work in forest science, which were prevented by the 1939 emergency. No attempt seems to have been made to reopen these discussions after the war, but members of the Committee who might be supposed to know the mind of the Agricultural Research Council thought that ARC would be glad to be relieved of any responsibility for grant-aiding fundamental research in forest science, and the Committee felt that the Forestry Commission should continue to administer its own research grants.

It was generally agreed, and indeed regarded as obvious, that the Commission must retain the capacity to do all strictly applied work. Robinson divided fundamental work into two categories: (a) basic work in the forest sciences not related to any specific problem, and (b) basic work having some bearing on applied forest research. He considered that the Commission should support the latter, but the former category was not the Commission's responsibility. It will be seen that the outlook had not changed materially since the early twenties and in particular there was still no single Research Council with 'forest science' in its terms of reference. How much this mattered would now be difficult to say. It certainly left the Commission, guided by its Advisory Committee, with the task of selling its research problems, and whilst it had always been good at doing this, it is likely that a Research Council would have drawn on a wider range of workers. But probably the main disadvantage lay in the complexity of the picture. By 1949 with the foundation of the Nature Conservancy, three Research Councils (DSIR, ARC, and the Conservancy) might support or be otherwise interested in items of research with forestry connections, and in addition the Commission maintained its own system of grant-aiding research of direct concern to forest problems.

However, in the early sixties the whole field of "civil science" and its organisation came under review. Firstly, the Advisory Committee on Scientific Policy set up a committee under Sir William Slater in 1961 to look into the special field of natural resources. This body recommended the grouping together of research activities of a broadly ecological nature under a new Research Council ("Slater Report", 1963). A Committee of wider terms of reference, the *Committee of Enquiry into the Organisation of Civil Science* (the "Trend Committee"), was appointed by the Prime Minister in 1962 to look into the whole field of civil science and its organisation ("Trend Report", 1963). The Trend Committee was in substantial agreement with the Slater Committee's recommendations for the organisation of research in the natural resources field, and amongst other important recommendations, advised that a new Research Council should be set up which would look after research in such subjects as terrestrial ecology, marine biology, hydrology, geological and soil surveys, and *long-term forestry research* [sic]. The Trend Committee's recommendations were accepted in the main by Government, and in 1965 the Natural Environment Research Council was established by Order in Council. The new body absorbed the Nature Conservancy, which, however, retained its identity and operated as a component part of the new council. This had five main committees, covering Ecology (the Nature Conservancy), Forestry and Woodland Research, Hydrology, Marine Biology, and Geological Survey. The Soils Surveys remained under the aegis of ARC.

NERC arrived rather late on the scene to have much influence on forest research during the period, but some comment will be made in the next chapter on the most recent developments in the new Council's work.

The essential feature of the early post-war period was the rapid expansion of the Commission's activities—especially the planting programme. It is difficult to find anything wholly new in the technical problems confronting Research, but there were many important shifts of emphasis, and a number of lines of work with discernible pre-war roots developed quickly to new projects or even to divisions of the research enterprise.

A few problems were directly connected with the wartime fellings. English broadleaved woodlands, which had by no means recovered from the first war fellings, were again devastated, and large acreages were left unstocked with anything of value, with only coppice regrowth or scrub of worthless species. This was a problem of greater significance to the private owner than to the Commission itself, but one which, as forest authority, the Commission was bound to investigate. There was also a marked deterioration in the condition of forest nurseries, and many forest operations such as drain maintenance had been neglected due to labour difficulties and shortage of supervision. Other examples could be given, but in most fields of research the incentive was the stage reached in pre-war work, and the developments followed naturally on the greater investment made subsequently.

As it happened, a good many lines of work had reached a critical stage and were open to new approaches and increased effort. This was very obviously the case in nursery nutrition, where Rayner's work had led to surprising and unexplained successes along a somewhat restricted line. In the afforestation of the important heathland and moorland types, very clear main lines of advance had been established, and the way was wide open for the development of machinery, cultural practices, and improvements in fertilisation.

In other fields, questions had been recognised, but little had been done because of shortage of workers and money. For instance, Robinson had for long been anxious to see more research done on forest soils, especially with regard to the changes taking place after afforestation. This became an important topic of basic research after the war when funds became available to support work at the appropriate institutes.

Another branch of study, forest genetics, which had been discussed before the war but dismissed as impracticable, was adopted after further debate largely due to the influence of work in other countries, notably Larsen's pioneer efforts in Denmark.

The most important policy decision, to establish a Forest Research Station, was by no means a new idea; the Commission had in fact been advised to set up an "institution" by a Cabinet Committee in 1920 (see page 5). As conceived in 1943, the new research station was not to be a very ambitious affair; it would provide a base for Silvicultural research in England and Wales, and would house specialists in Mensuration working on problems of production and management for the whole country; Ecology would also be represented, and most important, the Commission would have its own specialists in Pathology and Entomology at the station. Seed Testing was considered a likely development, and it was thought that a Photographic service should be developed.

Library facilities were (of course) considered essential. It was, however, made clear in discussions in the Research Advisory Committee that the Commission had no intention of embarking on research requiring "specialised knowledge and elaborate laboratory technique"; i.e. fundamental research would be left to the Universities and other proper institutions. There is a hint here of the confusion of ends and means which has so often bedevilled discussions on

kinds of research; sophisticated methods appearing more appropriate to fundamental than to applied research. The American Earl Stone once remarked that you could do pure research with a bulldozer and applied research with a microscope.

When the question of a Research Station was first discussed in the Advisory Committee, not all members were convinced of the need for it. One view was that it would lead to centralised or institutional research, whereas the strength of the Commission's research effort had lain in experimental work in the forests. Most members, however, were in favour of the station, but there was an interesting difference of opinion as to where it should be. The Commission put forward Alice Holt Lodge, a large house in the old Crown Forest of Alice Holt in Hampshire, close to Farnham, Surrey, and about 45 miles from London. The house had been a country residence, and during the war had served as a hospital. After inspecting the accommodation and facilities, most of the Committee considered that Alice Holt would be suitable for the purpose, but a strong minority view was that the station should be attached to one of the Universities, preferably Oxford, in order to keep the Commission's applied research in close touch with basic work.

At Oxford there would also have been the advantage of close contact with the Imperial (now Commonwealth) Forestry Bureau for documentation. Robinson, however, plainly felt that there was a danger of the Commission's own research becoming too academic and divorced from immediate field problems were the station to be based on a university. In the Commission's view it should not be difficult to maintain the necessary contacts with the Imperial (Commonwealth) Forestry Institute and with basic workers elsewhere, whilst maintaining a healthy degree of independence.

Policy decisions were made (explicitly or implicitly) on the opening of new sections or lines of work, and something will be said about the most important cases in the following chapter in tracing the main developments in forest research.

REFERENCES

"SLATER REPORT" (1963). Report of the Committee on Research into Natural Resources. In *Annual Report of the Advisory Council on Scientific Policy 1962–1963*, pp. 19-43. Cmnd. 2163. London: HMSO.

"TREND REPORT" (1963). *Committee of Enquiry into the Organisation of the Civil Service*. Cmnd. 2171. London: HMSO.

Chapter 11

ORGANISATION AND DEVELOPMENT OF RESEARCH

The changes in the organisation of the Forestry Commission consequent upon the passing of the Forestry Act of 1945 are fully described in Ryle's *Forest service* (Ryle, 1969). The period was one of very rapid expansion in the Commission's activities, and much thought had been given to the future role of forest research in the discussions leading to the production of the Command Paper *Post war forest policy* in 1943. The importance of forest research was recognised by the creation of a Directorate of Research and Education, in parallel with the territorial Directorates of England-and-Wales and Scotland which had replaced the old Assistant Commissioner charges. Appropriately, W. H. Guillebaud was the first to hold this rank.

The early post-war organisation of Research had, in growing, inevitably lost the primitive simplicity of pre-war days, when Robinson and Guillebaud were so close to the work that policy and execution could hardly be separated. Policy for Commission research and for support of external research now rested with the Director, who was responsible to the Director General. The territorial Directors had a formal opportunity to influence research objectives through the Technical Committee, and Robinson continued to take a keen personal interest in Research, through the meetings of the Research Advisory Committee and field inspections. His death in 1952 removed a dominant, personal influence (not least on Research) which could not be replaced by any organisational device.

Late in the period (in 1965) a further drastic reorganisation of the Forestry Commission (the principle change being to replace the territorial Directorates by a functional board structure at Commissioner level) placed the Directorate of Research under the permanent Commissioner for Forest and Estate Management. This should in theory have made the general adoption of advances based on research a simpler administrative process, and perhaps there are now signs that it has done so.

Alice Holt opened in 1946; that is staff began to trickle in to occupy the rather primitively furnished accommodation, and the station was hardly a going concern before the beginning of 1947. The promotion of Guillebaud to Director of Research and Education, stationed at London Headquarters, necessitated the appointment of a Chief Research Officer to supervise the enlarged Research Branch and to act as head of the new station. The first appointment to this post was M. V. Laurie, who had served as Silviculturist at the famous Indian Forest Service research centre at Dehra Dun. T. R. Peace, who had worked for the Commission by arrangement with the IFI for some years before the war, and had in fact served as a Commission District Officer during the war, took the post of Forest Pathologist. F. C. Hummel, who had worked in Uganda, opened the section of Forest Mensuration, the successor to the old Sample Plots Officer charge. H. S. Hanson from the Imperial Institute of Biological Control was appointed Forest Entomologist, and J. M. B. Brown, who had worked on entomological problems for the Commission at the Imperial Forestry Institute, joined the team at Alice Holt as Forest Ecologist after completing postgraduate studies in this subject at Oxford.

The Research Officers England-and-Wales and Scotland, R. G. Sanzen-Baker and J. A. B. Macdonald respectively, were re-titled Silviculturists, and thus the Silviculture Sections, which were the successors to the old general Research Branch, were the only sections to be divided on a territorial basis; the others (at least nominally) covered the whole country from Alice Holt. From 1946 Silviculturist (North) occupied office accommodation at Manor Place, Edinburgh; research staff in Scotland had previously worked from Scottish Headquarters in Drumsheugh Gardens, Edinburgh. The territorial boundaries were later modified to give Silviculturist (North) the northern counties of England in addition to Scotland, thus bringing most of the upland afforestation work together.

Alice Holt also served as a base for the Census of Woodlands (a Headquarters activity) under J. S. R. Chard. This was carried out between the years 1947 and 1949 as a matter of great urgency, and provided the essential statistical basis for much post-war policy.

In 1949 W. H. Guillebaud, the leading figure in forest research since the early twenties, left Research on promotion to Deputy Director General, being succeeded as Director by James Macdonald, who had much pre-war research experience. M. V. Laurie left the Commission in 1959 to take over the Chair of Forestry at Oxford on Sir Harry Champion's retirement; T. R. Peace succeeding to the post of Chief Research Officer. The numerous other changes in staff in the post-war period are shown below, but J. A. B. Macdonald's departure in 1951 on promotion to Conservator, South Scotland, must be mentioned, since this closed a record of nearly thirty years in Scottish research, during which he became the leading expert in upland afforestation.

Research

Directors

W. H. Guillebaud 1945–1949 (Deputy Director General of the Commission 1949–1953)

J. Macdonald 1949–1962 (Deputy Director General of the Commission 1961–1962)

A. Watt 1963–1965 (Forestry Commissioner 1945–1969)

J. R. Thom 1965–1968

G. D. Holmes 1968–1973 (Forestry Commissioner 1973–)

D. R. Johnston 1973–

Chief Research Officers

M. V. Laurie 1946–1959

T. R. Peace 1959–1962

R. F. Wood 1962–1967

D. H. Phillips (CRO, South) 1967–

B. W. Holtam (CRO, North) 1968–

Heads of Sections

Seed　G. M. Buszewicz 1963–

Silviculture South　R. F. Wood 1947–1962, R. M. G. Semple 1962–1970, J. R. Aldhous 1970–

Silviculture North　J. A. B. Macdonald 1948–1951, M. V. Edwards 1951–1966, R. Lines (acting) 1964–1966, D. T. Seal 1966–1970, R. M. G. Semple 1970–

Ecology　J. M. B. Brown 1964–　　(Retired in 1972)

Soils　W. O. Binns 1963–

Genetics　J. D. Matthews 1947–1963, R. Faulkner 1963–

Pathology　T. R. Peace 1947–1959, J. S. Murray 1959–1961, R. G. Pawsey 1961, D. H. Phillips 1961–1967, D. A. Burdekin 1967–

Entomology　H. S. Hanson 1947–1951, M. Crooke 1952–1960, D. Bevan 1960–

(*Mammals and Birds* Sub-Section: Miss J. J. Rowe 1961–) (Now *Wild Life* Section)

Physiology　I. D. J. Phillips 1968–1969, K. A. Longman 1970–　　(Resigned in 1972)

Statistics　J. N. R. Jeffers 1956–1968, R. S. Howell 1968–

Publications　H. L. Edlin 1945–　　(Also Librarian 1962–1966)

Library and Documentation　G. D. Kitchingman 1947–1960, R. M. G. Semple 1960–1962, H. L. Edlin (part-time) 1962–1966, O. N. Blatchford (Research Information) 1966–

Photography　I. A. Anderson 1949–

Management Services

Directors

F. C. Hummel (Conservator) 1966–1968

J. Q. Williamson 1968–1970

D. R. Johnston 1970–73

Deputy Director

D. R. Johnston 1968–1970

Heads of Branches

Mensuration　F. C. Hummel 1946–1950

Census　J. S. R. Chard 1947–1950

Mensuration and Census　F. C. Hummel 1950–1956

Forest Management (*embracing Planning and Economics, Mensuration and Census*)　F. C. Hummel 1956–1961, D. R. Johnston 1961–1963

Planning and Economics　D. R. Johnson 1963–1966, J. A. Spencer 1966–1968, D. R. Johnston 1968–1970, A. J. Grayson 1970–

Field Surveys　L. M. Simpson 1970–

Work Study　J. W. L. Zehetmayr 1954–1964, L. C. Troup 1964–1971, I. A. D. Grant 1971–

Machinery Research　R. G. Shaw 1949–1965, A. J. Cole 1966–1967, R. B. Ross (Work Study) 1967–

Utilisation

Utilisation Development　E. G. Richards 1950–1966, B. W. Holtam 1966–1968, J. R. Aaron 1968–

From the structure of the branch outlined above, a number of developments took place in the post-war period, and these will be reviewed briefly. The first new Section to be added was Forest Genetics, which was started by the appointment of J. D. Matthews in 1947. The possibilities of tree breeding had been discussed in the Research Advisory Committee as early as 1933, when a paper by Nicolai of Dantzig making proposals for international work was considered. At this time the scheme was considered much too ambitious. However, advances in provenance and interest in Continental breeding work encouraged further discussion of the topic, and in 1946 Robinson asked Sir Harry Champion to chair a sub-Committee of the Research Advisory Committee to look into the possibilities of forest genetics in its application to British Forestry. Sir Harry Champion's Committee reported in favour of taking up this work, and the Commission accepted the recommendation. But it is not without interest that one of the arguments against the proposal was that more was to be gained by attention to seed supply at the species and provenance level. Whilst this did not logically exclude tree breeding, it is certainly the case that the organisation of seed supply remained weak over much of the post-war period, and a good deal of productive capacity was lost thereby.

The Mensuration Section under F. C. Hummel, successor to the old Sample Plots Officer, evolved beyond recognition during the period. A sub-Section of Statistics grew up inside it under J. N. R. Jeffers and developed a service to the Branch. In 1956 Statistics became an independent Section concentrating on the design and analysis of experiments. At the same time the Mensuration Section was retitled Forest Management, and expanded from the original role of studying growth and yield to assume responsibility for Working Plans, adding at the same time a new discipline, Forest Economics, with the appointment of A. J. Grayson. Management had also taken over the Headquarters activity of Census Woodlands. In 1960 the Management Section left Research and became a Headquarters section. Since the greater part of its work was now a common service to the Commission, this was not illogical, but it would have been better if at this time the opportunity had been taken to disengage the purely research component in Mensuration and leave this in the Research Branch. Later in 1966 Management, Work Study, and Machinery Research were incorporated in Management Services Division.

The two Silvicultural Sections, North and South, expanded greatly over the period as the range of work increased. The small Ecological Section under J. M. B. Brown remained linked to Silviculture (South) till 1964, when it became independent. In the nineteen-fifties the need for soil analytical work became pressing, since it was not always convenient nor practicable to farm out *ad hoc* soil investigations. A small sub-section allied to Ecology was established under W. H. Hinson in 1955. Later under W. O. Binns the Section also became autonomous. Work on seed analysis and pre-treatment had been assuming increasing importance since the establishment of the Research Station, and by 1963 the Forest Seed Section which had grown up under G. Buszewicz inside Silviculture (South) became independent. By the early fifties it became apparent that Alice Holt Lodge could not continue to satisfy requirements for space, in spite of a proliferation of ex-army huts round about the house. The laboratories were makeshift conversions of domestic accommodation and very cramped; the new wing was planned by Laurie (architects Cowper and Poole) to provide adequate modern laboratories for all the sections in need of special facilities. It also included refrigerated seed stores to replace the very primitive shed and cellar stores at Santon Downham (near Brandon, Suffolk) and Grizedale (Lancashire), and other Conservancy stores. The new building was opened in 1959.

As well as these developments in Research proper, the post-war period saw the addition of certain specialised sections to the Headquarters organisation, which were concerned, at least in part, with research activities. The first of these, Mechanical Development, dates from 1949 with the appointment of Col. R. G. Shaw. In 1950 a Utilisation Development Section was formed under E. G. Richards, one of its more important roles being to liaise with Princes Risborough and co-ordinate research on homegrown timbers with the Forest Products Research Laboratory, Princes Risborough. And in 1954 J. W. L.

Zehetmayr left Silviculture (North) to start a new Section in Work Study. (It is of interest to recall that the Commission had made some approach to this class of work with O. J. Sangar in the early twenties). These Sections did not report to the Directors of Research, but to Commission Headquarters. With the re-organisation of 1965, and the division of the Commission's activities into three broad functions under permanent Commissioners (Administration and Finance, Forest and Estate Management, and Harvesting and Marketing) whilst the main Research Branch found itself in Forest and Estate Management, Utilisation research passed to Harvesting and Marketing, and Work Study joined Management Services in Administration and Finance.

Whilst Alice Holt had considerable advantages as a Forest Research centre, it had also some drawbacks. For one thing, with the best intentions, it was difficult for specialist staff to devote as much attention to Scotland and the north generally as was justified by the increasing relative importance of forestry in these regions. Even had this not been so, changes in approach in the main experimental section, Silviculture, called for some sort of laboratory facilities, and at Manor Place and (latterly) Sighthill in Edinburgh, Scottish research had only the bare minimum office accommodation. By the mid-sixties attention to Scottish requirements was plainly overdue. One important step had already been taken when J. D. Matthews left the Commission in 1963 for the Chair in Aberdeen when it was decided to base the Genetics Section at Edinburgh under R. Faulkner, leaving a deputy at Alice Holt to look after the work in the south. In 1967 it was decided to build a new Research Station at Edinburgh, and a site was provided by the kindness of the Board of the Edinburgh Centre of Rural Economy at their Bush Estate, near Roslin, Midlothian. As the building was completed in 1970, developments at Edinburgh fall outside the period reviewed, but it should be said that the new station houses Silviculture (North); a new Section in Physiology; Forest Genetics; and representatives of the Pathology, Entomology, Soils, and Statistics Sections. Its situation, in a campus of other research establishments (Hill Farming, Plant Breeding, Agricultural Engineering, etc.), and close to Edinburgh with a strong research background at the University Department of Forestry and Natural Resources, is obviously most favourable.

The research programme has continued to be initiated chiefly inside Research; that is, it has been the Research interpretation of the Commission's needs rather than a list of projects "sold" to Research by the Headquarters organisation. There have been various changes in drawing up and reviewing the programme. It has always been the main item at the annual research conference, and has always been presented to the Research Advisory Committee for general comment. The Committee has been influential in encouraging certain lines of work, e.g. nursery nutrition; forest soils; genetics etc., and has also been of great value in grant aiding through the wide scientific contacts of its members. Whilst that body still existed, the Commission's Technical Committee made the primary official scan of the programme. Various methods of consulting field staff have been adopted, by far the most valuable being meetings for discussion on the ground.

Generally speaking, the main development in the programme itself has been from a detailed list of items of work (which became increasingly cumbersome with the growth of the Branch), to a more convenient summarised format setting out the directions of work. This has been facilitated by the more or less general introduction of the project as a formal basis for planning in the mid-sixties. In some fields (Work Study is a good example) the term has immediate significance, and has been applied from the outset. Elsewhere it has not always been easy to demarcate projects from a wide field of inter-related studies, but even where the definition is somewhat arbitrary, it has been useful in forcing attention to broad objectives, which do not necessarily appear in individual experimental plans.

Advances in the application of statistical theory have improved the experimental work and other enquiries undertaken by all sections of research during the post-war period. One important early influence was the demonstration by E. M. Crowther of the value of the factorial designs (largely due to F. Yates) in his experiments on nursery nutrition (Yates, 1937). The setting up of a general statistical service, first inside Mensuration, but from 1956 an independent Section, has ensured that the best methods were available to all workers.

Jeffers' term of reference were to advice on design, to analyse the results of experiments and surveys, and to research into the application of statistical methods in forest research and management. At this time a fully fledged statistical service was an unusual feature in a forest research outfit, and not by any means general in comparably sized establishments in other fields. Most of the special problems in forest research stem from the long-term nature of the crop, the relatively large size of plots and the heterogeneity of forest sites. Hence, there has been much use of techniques of covariance in accounting for original differences in experimental material. The introduction of new designs such as incomplete randomised blocks were of great assistance in keeping down the number of plots in experiments with large numbers of treatments (as in provenance studies); and other devices such as partial replication in factorial experi-

ments proved useful in forest nutritional studies. A great deal of attention was paid to sampling in experimental assessment and in surveys of all kinds. Schemes of stratified random sampling were adopted in many of the more important surveys, as in those concerned with the joint studies on timber properties with Princes Risborough. Jeffers also had an early interest in sequential sampling, and applied it in one of the major investigations on the strength of home-grown pitprops. One of his more important research topics was on growth curves; mathematical treatment of these has applications in mensuration and forecasting (Jeffers, 1959).

In the mid-fifties it was becoming clear that the weight of routine calculations in analyses and in mensuration and other topics was becoming too much for desk machines; also some of the new statistical techniques were hardly practicable by these means. At the time electronic computers had reached an advanced state of development, and third generation types were becoming available. Jeffers had access to the Ferranti Pegasus at Farnborough through the kindness of the Royal Aircraft Establishment, and programmed many analyses and other calculations for this machine, and later for the related Ferranti Sirius at the London Computing Centre. By 1960 virtually all the major calculations undertaken by his section were being undertaken on hired time on these computers. The advantages of acquiring a computer at this time were strong; for one thing, a good deal of development work in methods was necessary, and the relatively short period of hired time available were inconvenient for this, and usually had to be filled with routine work. In 1963 the Commission acquired its first computer, a Ferranti Sirius, which was installed at Alice Holt and performed excellent service over the rest of the period.

There are obvious advantages in getting numerical data onto the computer as directly as possible; time saving, avoidance of copying errors, etc. One method entails the "reading" of marks on specially printed forms; this is the basis of the English Electric Lector which was used experimentally for the field data forms of the 1947–49 Census. The Commission was kindly permitted to try out the method on a machine owned by a well-known firm of confectioners; a terse diary entry "visit to Mars" caused some consternation in Establishments. Portable punch machines are now thought to have wider application in capturing field data. One intriguing piece of ancillary equipment which has been added is the Calcomp Graph Plotter, which plots points directly from formulae programmed through the computer.

Without embarking on any discussion of the impact of the modern computer, two points may be made briefly. Firstly there is the very obvious one that it has removed the boring and lengthy slog in numerous routine calculations; in forestry some of the most important gains at this level have been in mensuration and enumeration. Secondly there are all the types of calculation which it is hardly practicable to attempt without it; analyses of large and complicated experiments, multivariate analyses, linear programming etc. In forest experimental work, statistical advances plus the computer (the two are hardly separable) have encouraged the laying down of larger and more complex experiments and the measurement of more variables, rather than an increase in the number of experiments; this usually means more information for a given effort. However, nothing illustrates Parkinson's Law better than the computer in a research establishment; it is promptly submerged with work. One of the heads of department at Rothamsted, on being asked how their latest large machine was getting on, replied "quite well, it had only delayed his routine analyses two months". Whether an establishment should possess its own computer or have access to some larger central installation is outside our scope; whatever the system, there are advantages in as many workers as possible being able to program and put their own material through the computer. Jeffers actively encouraged this, and Alice Holt provided training in statistical method and computation for many visiting students, in addition to Commission staff.

Something should be said here about developments in grant-aided and cooperative research over the post-war period. Very much more financial support was given than in the pre-war period, and the scope of the investigations which were aided by the Commission widened considerably. Some of the investigations will be mentioned in succeeding chapters, though not adequately in view of the volume of the work. The arrangements for grant-aiding varied. The commonest circumstance was a grant for a three-year period for a specific investigation to support a research worker at a University or other institution in respect of laboratory assistance, special equipment, travel expenses etc; the grant being often renewed for a further period if required. But there were some quite small grants for travel expenses alone to encourage University staff to look at particular problems in Commission forests. Some of the "grants", though spoken of as such, were rather semi-permanent arrangements with institutions to follow particular programmes of work, such as those with Rothamsted Experimental Station on nursery nutrition, and with the Macaulay Institute on forest soils generally. The Commission also paid the salaries (through the institutions concerned) of individual workers on forest soils—an arrangement close to outstationing. Truly co-operative research, though theoretically attractive, is often difficult to arrange—a programme of work

usually has to have one boss. However, there were a number of cases in which projects were very profitably shared; the joint programme on nursery nutrition with Rothamsted being a most successful example. A comprehensive joint programme of research into the properties of home-grown timbers was started in 1958. The Forest Products Research Laboratory had done some work on home-grown timbers before and during the war (including important investigations on pitprops), but the beginning of production on a rapidly rising scale from the Commission's own plantations of the twenties and thirties was an incentive towards a more radical approach. The programme, of which more will be said in Chapter 27, was directed by a joint Committee of the Forest Products Research Laboratory and the Forestry Commission.

During the period, the Nature Conservancy developed its own programme of woodland research, some items of which (especially work on the effects of tree crops on the site) were of direct interest to the Commission. There was a good deal of useful informal co-operation, and at least one important joint project, the Gisburn experiment at Bowland Forest, Yorkshire, on the long-term effects of conifers and broad-leaved species.

The establishment of the Natural Environment Research Council was mentioned in the preceding chapter. The Commission was represented from the outset on its appropriate Committee; the Forestry and Woodlands Research Committee. By its terms of reference, NERC became the proper agency for the support of basic forest research, and as the Research Council organised its grant system, the Commission shifted its own grants towards investigations of immediate practical value and away from anything of long-term speculative nature. The new Council has (in addition to its supporting and co-ordinating role) the powers to undertake forest research on its own account, and at the time of writing the Council has set up a research unit at Edinburgh University, with a proposed new research station (The Institute of Tree Biology) planned to pursue certain types of work which require continuity and are thus not wholly suited to units based on universities. FAWRC has considered that the main gaps in the present research pattern are basic work in tree physiology and genetics. The Commission had already decided to set up a Section in Tree Physiology based on Edinburgh, and it is obvious that the next decade will see a process of adjustment between Commission research and that conducted or supported by the Council. In many ways it is to be regretted that a Research Council concerned with forest science has arrived on the scene so late in the day. Had it appeared in the late forties, the adjustment would have been much easier, but having had to "go it alone" Commission work now contains such an amalgam of basic and applied work in so many fields that separation on the lines of agricultural research will be difficult, if indeed desirable.

REFERENCES

JEFFERS, J. N. R. (1959). *Experimental design and analysis in forest research*. Stockholm: Almqvist & Wiksell.

RYLE, G. B. (1969). *Forest service: The first 45 years of the Forestry Commission*. Newton Abbot, Devon: David & Charles.

YATES, F. (1937). *The design and analysis of factorial experiments*. Tech. Commun. Imp. Bur. Soil Science 35.

Chapter 12

FOREST TREE SEED

As mentioned in Part I, pre-war work on seed had been confined to pre-treatment to accelerate germination, or relieve dormancy in certain species. Storage practices (duration or condition) do not seem to have posed any particular questions. Germination tests followed standard agricultural practice, and though not regarded as wholly satisfactory, appear to have been reasonably adequate.

Post-war research on seed stemmed directly from the old interest in pre-treatment, but the first, crude (bench and sink!) laboratory facilities at Alice Holt made it possible to do some work on test methods in order to assess the results of pre-treatment. An important stimulus was the published work of the German G. Lakon on biochemical staining methods of germination testing, especially his development of a process using a salt of tetrazolium (Lakon, 1947). This was initially attractive because of the speed at which it could be carried out, and the modest facilities required. Work on tetrazolium is mentioned by G. D. Holmes in the (unpublished) research report for 1948, and he was also concerned at this time with work on pre-treatment (soaking in water and acid pre-treatment), and had begun the first work on storing methods. At this stage, work on seed was treated as an adjunct to nursery research (logically enough), which was Holmes' main concern.

Pre-treatment as a *nursery* practice did not advance much in the early post-war period, and interest shifted to the allied subject of the physical conditions in germination testing. This was partly a matter of expediency, since the expanded nursery programme encouraged the Commission to carry out its own seed testing, and it was clear from an early stage that forest tree seed required some modifications of standard agricultural practice. The technique of "pre-chilling" (exposure to 3–5°C on wet substrata, i.e. "naked stratification") proved to be of value in speeding up the germination of several conifer species. Varying amounts of dormancy in some species required two or more simultaneous tests to detect the degree of dormancy. This "duplicate" testing was introduced into the testing rules of the International Tree Seed Association which the Laboratory joined in 1952. On the whole, dormancy problems have not rated of the first importance with the principal species, and some proved of less significance with improvements in storage conditions.

Laboratory test methods were improved and standardised for purity, moisture content (including the toluene method); sampling and mixing; and for germination. Quick methods of germination testing were X-ray, excised embryo and biochemical stains including indigo-carmine, sodium biselenite and finally tetrazolium. It was quickly realised that though the individual test was speedy a large number of samples would require an extravagant laboratory staff. Hence "actual" germination methods were adopted, with tetrazolium used to assess the viability of dormant seed failing to germinate within the standard period. Buszewicz and Holmes (1957) published an account of their experience with this method. (Plates 30 and 31.)

During the first year or so of routine testing, the laboratory tested only the seed to be sown in Research experiments, plus some late-arriving consignments for production nurseries, but by 1951 it was sufficiently well established to be licensed by the Ministry of Agriculture as a Private Seed Testing Station, and since then all seed for use or sale by the Commission has been tested here; the laboratory being under the charge of Gwidon Buszewicz. Hence starting as a purely research activity inside the Silvicultural Section, Seed took on a predominantly service function. Since 1963 it has ranked as a separate Section.

Work on long-term storage was pursued during the late forties and early fifties. It was realised at this time that much seed was being stored under very makeshift conditions, and also that temperature and moisture control were the key factors. The construction of the new wing at Alice Holt, which was completed in 1959, provided well-designed seed laboratories and ample refrigerated storage space for at least two years' reserve of seed. Facilities for processing, drying, mixing etc. were also provided, and from 1959 Alice Holt became the Commission's centre for storage and supply as well as testing. This fitted in very conveniently with seed testing, since sampling (which was given a good deal of attention), processing and testing could be programmed to suit the laboratory. Taking on the seed "business', however, involved a great deal of development work.

The early years of the seed laboratory coincided with a period of extreme activity in the Commission's nurseries, with much emphasis on production. This was a great stimulus to improve not only the test procedures but to find the best expression of the vital statistics for prescribing sowing density. The difficulty here of course is that the tester can only say what a seed lot will do under optimum conditions; nurseries differ, and he cannot prescribe for the differences. One administrative approach in this period was to determine germination and survival factors for individual nurseries. From the research side, a good deal of attention was paid to germination energy, or the *rate* of germination as against the proportion of a seed sample germinating in a standard time (germination capacity). There was reason to suspect that germinative energy might prove a better basis for sowing prescriptions than germination capacity under poorer nursery conditions. Two things, however, combined to reduce the significance of these approaches. Firstly, the great improvement in storage conditions in the new stores at Alice Holt diminished the range of germinative energy in seed lots, and secondly the shift of sowing towards the new, acid, heathland nurseries provided much better conditions for germination and survival. Hence, sowing prescriptions based on standard germination tests became more meaningful and reliable. A useful innovation was the expression "effective pound", i.e. the number of viable seed to be expected per pound of any particular seed lot, since this took in variations in seed weight and made nursery calculations easier.

By 1962 the Commission decided to erect a new central cone extractory for home collections at Alice Holt. A good deal of study of the most up-to-date methods and some development work was required in equipping this. The extractory first functioned in 1965. Home seed collection required a number of studies of a strictly practical nature, mostly concerned with methods of assessing yield from crops and individual cones, the work being shared between the Seed and Genetics Sections.

What we may call the main line of seed work—everything concerned with storage, testing and prescription—culminated in the mid-fifties. Part research, part development, it had been carried through extremely fast. Since it coincided with a period of notable improvements in nursery management, of active research in the nurseries with marked advances in knowledge and technique, it is difficult to estimate the value of the seed work by itself. What can be said, however, is that it had a large share in the very great economies in forest tree seed which became increasingly apparent in the sixties. From 1957 to 1967 the Commission's consumption of conifer seed dropped very steadily irrespective of seedling production; some 14,000 lb less seed being used in the latter year for a similar outturn of seedlings. Merely to give the order of the saving, such an amount of seed might cost £150,000 to purchase, but there are numerous concomitant benefits in seed economy; for instance seed of high genetic quality goes further, with long-term benefits to the plantations.

There is no sharp distinction between laboratory and nursery work on seed, and some lines of work of nursery significance fed back to the laboratory. For example, the very old practice of dressing seed with red lead (mentioned in Part I, Chapter 8, page 30) requires the moistening of the seed for adherence of the colourant. It was known that the red lead conferred no particular advantage other than showing up the seed, and moistening was considered objectionable if for any reason sowing had to be delayed. In the late sixties attention was paid to dyes which could be applied with oil as adhesive material, and by 1968 'Waxoline' type dyes had given favourable indications. This was one example of the possibility of central processing of seed and by the end of the period this seemed the most profitable line of advance. The pre-treated, pre-sown unit, is by no means an impossibility, since it has already found some place in horticulture.

Work on seed pathology is perhaps best mentioned under Nurseries, since the main line of work was followed at Rothamsted as an adjunct to the Nursery Nutritional sub-Committee's programme.

REFERENCES

BUSZEWICZ, G. (1962). The longevity of beechnuts in relation to storage conditions. *Rep. Forest Res., Lond.* 1961, 117–126.

BUSZEWICZ, G., and HOLMES, G. D. (1957). Seven years' seed testing experience with the tetrazolium viability test for conifer species. *Rep. Forest Res., Lond.*, 142–151.

HOLMES, G. D., and BUSZEWICZ, G. (1956). Longevity of acorns with several storage methods. *Rep. Forest Res., Lond.* 1955, 88–94.

HOLMES, G. D., and BUSZEWICZ G. (1958). The storage of seed of temperate forest tree species. *For. Abstr.* **19**, 313–322 and 455–476.

HOLMES, G. D., and BUSZEWICZ, G. (1959). The assessment of seed sowing value. *Q. Jl For.* **53** (3), 220–227.

LAKON, G. (1947). The topographical tetrazolium method for determining the germinating capacity of seeds. *Pl. Physiol.* **22** (3), 389–394.

Chapter 13

NURSERY INVESTIGATIONS: NUTRITION

Chapter 3 in Part I left the story at the opening of the first heathland nurseries at Wareham, Dorset, and Harwood Dale in Yorkshire, as an outcome of Dr Margaret Rayner's studies. Her composts had given excellent results on acid heathland soils, but the mode of action was not understood, and the composts were expensive and laborious to prepare. The Research Advisory Committee discussed nursery nutrition at Newcastleton in July 1944, and it was at this meeting that Lord Robinson accepted the advice of G. V. Jacks and Prof. F. T. Brooks to call for the assistance of Dr E. M. Crowther of Rothamsted. The arrangement made with Sir William Ogg, then Director of Rothamsted, was a straightforward financial grant from the Commission, but the programme was directed by a sub-committee of the Research Advisory Committee; the "Sub-Committee on Nutritional Problems in Tree Nurseries" (NNC for short).

Prof. F. T. Brooks of Cambridge held the chair of the sub-Committee till his death in 1952, when Prof. H. M. Steven took over and held the post till the sub-Committee dissolved itself in 1962, having seen the investigations through to a satisfactory stage. The arrangement was a curious one, but it worked extremely well in practice. The Commission gave a high priority to the NNC programme in Research staff and nursery facilities; for a number of years Kennington Research Nursery at Oxford under W. G. Gray, and Wareham Research Nursery, Dorset, under E. Fancy, were largely given over to this work. One of the Alice Holt Silvicultural staff spent much of his time on arrangements for the programme.

Dr Rayner remained active till her death in 1949; in the early post-war period her practical interests had shifted from the forest to the nursery, and part of Wareham Nursery was given over to large-scale experiments on the use of composts and mulches in the maintance of nursery fertility. With her death, this side of the work became less active, though certain of her prescriptions were adopted as the standards of organic manuring in comparative experiments with soluble fertilisers, and demonstrations of her methods were maintained for many years. Dr Ida Levisohn, who had assisted Dr Rayner for some time in her mycorrhizal studies, continued this work under Commission grant, specialising latterly in the stimuli to growth brought about by rhizosphere influences of fungi prior to the actual formation of mycorrhizal associations.

Crowther found it convenient to carry out the bulk of his experimental work in the south, i.e. in the Silviculture (South) region, the principal research nurseries being supplemented by sites in a number of "production" nurseries within reach of the Kennington and Wareham staff. Most of his forest extension experiments were performed in Wales on upland *Calluna* moor.

The NNC also exercised some oversight over Scottish work on nursery nutrition, Dr A. B. Stewart being one of the most active members of the sub-Committee; the Macaulay Institute, however, advised and collaborated, and did not *direct* the experimental programme. The programme in Scotland was not so large, and was not concentrated in Research nurseries, but spread over a number of the important production nurseries.

No attempt will be made here to summarise Blanche Benzian's *Experiments on nutrition problems in forest nurseries* (Benzian, 1965), which is the comprehensive account of the Rothamsted/Forestry Commission collaboration in the main nutritional field, and in certain ancillary topics, over the period 1945–62. Brief comment must, however, be made on the development and impact of the work.

It will be recalled that the situation had two outstanding features: (a) the success of composts in acid soils, and (b) with few exceptions, the failure of conifer seedlings to respond to *any* form of manuring in the older nurseries of agricultural soil type (the so-called established nurseries).

Two vital advances were made in the first few seasons of the Rothamsted programme. Crowther suspected that soil reaction was concerned in the poor performance of conifer sowings in Commission nurseries, and his first analytical surveys of production nurseries in England and Wales supported this. By 1947 calcareous seedbed covering materials and additions of lime through composting were outlawed, and experimental comparisons of covering materials subsequently confirmed the importance of this decision. The other great advance was the demonstration that conifer seedlings on acid soils responded satisfactorily to soluble fertilisers—much as other crop plants in fact—and that such fertilisers could be applied to the seedbed before sowing without detrimental results which were expected at this time. The Nursery Nutrition sub-Committee was able to report in 1947 that seedlings had been raised in acid nurseries with NPK fertilisers which were of apparently equal quality to those raised on Rayner-type composts on the same soils.

These two findings provided the keys to all subsequent advances in nursery nutrition. The main line of advance centred on naturally acid soils, uncomplicated by pathological factors. The heathland nursery had shown its first successes with compost manuring and work on composts and the processes of composting continued for some years; but by the early fifties it had become apparent that raw hopwaste (the principal substance used in composting) was as good as composted hopwaste, and from this time raw hopwaste became the standard organic manure in Commission nurseries. It was possible to make manurial prescriptions after a few seasons' work in Crowther's programme, and the general practice of applying PK fertilisers and raw hops to beds in advance of sowing, with N topdressings to growing seedlings, was established at an early date.

Much work was required on rates, formulations, times and methods of application of fertilisers, and on elements other than NP and K. No details will be given here. As the whole programme had started in an atmosphere of the special values of organic manuring, a great deal of attention was given to comparisons of organics and fertilisers in the short term, and working up to long-term fertility experiments. It proved remarkably difficult to demonstrate any benefits special to organic manures other than those attributable to their nutrient contents (including, in certain cases, elements other than N, P and K —such as Mg and Cu), and it required some twelve seasons of continuous seedling cropping before the performance of seedbeds manured purely with fertilisers began to fall regularly below those receiving hopwaste in addition. Nor was it possible in the Rothamsted programme to show any advantages from rotations, fallows or cover crops.

Long-term fertility experiments in Scotland differed slightly from the results at Kennington and Wareham in that the straight fertiliser treatments did not maintain their lead over hopwaste for so long. One important and quite general result, in all comparisons between soluble fertilisers and organic manures in seedbeds, was that the yields of seedlings were greater where organic manures were not applied.

Scottish work in the post-war period followed the general lines of the NNC programme with differences of emphasis. The advisory and analytical services of the Macaulay Institute encouraged individual nursery manuring prescriptions based on analyses, as against the block prescriptions which were adopted in the south. The distinction between manuring the soil and manuring the crop was more apparent than real, but what was undoubtedly of value in Scotland was the intimate knowledge of production nurseries built up by the Macaulay/Research Branch collaboration.

One line which was followed for some years in Scotland, but not in the south, was the placement of fertilisers. This was suggested by Dr Stewart. The idea was that if forest nurseries moved towards full mechanisation, band or drill sowing would have advantages over broadcast, and placement of fertilisers below the drill would be the obvious method. The technique proved quite satisfactory so far as nutrition was concerned, but the yields from drills or bands never equalled those from broadcast sowings.

In the late sixties nursery nutritional work in Scotland concentrated on the slow release fertilisers magnesium ammonium phosphate, potassium metaphosphate, and various forms of nitrogen for incorporation in the seedbed.

The transplant stage shared in nutritional studies and the old technique of the "extension" experiment was used to test the residual effects of seedbed manuring on growth in the lines, and of both seedbed and transplant manuring on growth in the forest. Such experiments usually included fresh applications of fertilisers, and one by-product of this stage of the work was the dismissal of the old fear of soluble forms of phosphatic manures, for few if any deaths were observed from such cause over many seasons' work. The difficult criterion of quality (as against size) of nursery stock received a good deal of attention at all times, for it was thought likely that highly contrasting nursery regimes—organic versus inorganic—might produce plants of the same size but different survival values. In the event, there appeared to be little in this; in fact the Research nursery $1+1$ Sitka spruce transplant, raised by whatever means, proved to be a nearly indestructible object. Another more specific anxiety was that full nutrition might make plants soft and liable to damage by frost or exposure in the forest. The results of many extension experiments dismissed this as a general proposition; though applying nitrogen very late to certain conifers may prolong their growth and render them susceptible to autumn frosts. But though fears of over-feeding in the nursery were fairly easily dismissed, it proved very difficult to separate the effects of size of plant from those of nutrient contents on behaviour in the forest, and this was first done satisfactorily late in the programme, when it became possible to fortify plants with K and N after their growth had stopped. Benzian's work on this line has begun to show that there may be benefits to early growth and in resistance to frost and exposure from high nutrient content in the nursery plant (Benzian and Freeman, 1967). There is an interesting link here with Forestry Commission work on fertilisation on exposed sites, where N nutrition (especially) has been shown to improve the resistance of the young tree to exposure.

The agricultural soil type nurseries presented a much more complex set of problems than the new heathlands. They varied in soil texture, reaction and in age. Many, including some of the largest, were sited on soils which were too heavy for ease of cultivation. Most of the older ones exhibited what was termed "Sitka sickness"; i.e. the inability to produce seedlings of reasonable size with the concomitant troubles of frost lift and low yield. Weed populations were high after the difficult war years.

Nutrition, even in the broadest sense, was only one aspect of the problems of the established nursery. The two principal ameliorative treatments applied to 'Sitka sick' nurseries were partial soil sterilisation and acidification. Research in Scotland was first in the field with sterilisation by steam and formaldehyde; the early experiments at Newton Nursery, Moray, in 1945 being a marked success. Such treatments were derived from glasshouse practice. Alice Holt followed Scotland in this line, and for a few years the Commission developed the steaming technique, modifying glasshouse methods to suit open beds. In England, Crowther laid down experiments over steamed plots for a few seasons. Steaming, though highly successful, was costly and inconvenient, and was abandoned in favour of formalin by 1953, and in England further enquiries into partial soil sterilisation were then left to the Rothamsted programme. Partial sterilisation by formalin was developed to user standards, and has been practised in a few Scottish nurseries for some years. In the Rothamsted programme many other sterilants were investigated; one practical target being to reduce the volume of water applied to seedbeds. Chloropicrin proved most promising amongst the fumigants. However, with the shift of conifer sowings to the heathlands, partial soil sterilisation ceased to be of practical importance, and became of more value as a diagnostic tool in soil microbiological studies. Crowther exerimented on acidification with sulphuric acid, aluminium sulphate etc., from the beginning of his programme, such work being confined to nurseries without great excess of free lime. The approach was successful, but probably the greatest benefits were diagnostic; and in the longer term, in providing a basis for the control of nursery reaction through the choice of nitrogen fertilisers. Plates 32 and 35 show steam sterilisation.

Work on partial sterilisation drew in a number of biologists, from Rothamsted and other institutions specialising in different groups of soil micro-organisms. Janet Brind (1965) of Rothamsted studied the differential effects of partial soil sterilisation on soil bacteria and fungi, the short-term effect being a great reduction of the latter coupled with an increase in the bacteria—particularly denitrifiers. This provided confirmation of Crowther's theory that conifers such as Sitka spruce might prefer their nitrogen in the ammonium rather than the nitrate form. D. M.

Griffin (1955–57; 1965) of Cambridge concentrated on the fungi causing death and root damage in young seedlings: he ascribed the most important role to *Pythium ultimum.* J. B. Goodey (1965) of Rothamsted examined the special case of the nematode *Hoplolaimus,* which was shown to be playing a major part in root damage in Ringwood Nursery, Hampshire. Subsequently, the lines of work started by Brind and Griffin were further developed by F. T. Last and R. A. Ram Reddy (1963; 1964) at Rothamsted, and latterly G. A. Salt (1964–70) made a special study of soil pathogens and early germination troubles, which led him to the isolation of a seed-borne fungus capable of causing significant pre-emergence losses, especially where seed lay inactive at low temperatures. Reference to these later studies will be found in the *Report on forest research* from 1962 onwards.

The gains from work in this general field in the period were large and widely spread. The most obvious were related to the rate of production of plantable stock, for under favourable nursery conditions it became practicable to obtain high yields of usable one-year seedlings capable of producing plantable stock after a year in the transplant lines. This offered the prospect of large economies in nursery space and tending costs related to time and area, but to achieve these economies a great deal of reorganisation was required, the principal change being the shift of sowing programmes to new acid soil nurseries taken in from heath, or pine woodland sited on heathy types. Many of the old nurseries had to be maintained and the policy was to use them for transplanting seedlings used in the new nurseries. However, the eventual economies in nursery ground were very substantial; by 1970 for instance the Commission had reduced its nursery space to about 1,100 acres from the 2,200 acres in use in 1950, when the planting programme was of much the same order.

The biggest practical gains stemmed from work in heathlands and acid soils of agricultural type, and in retrospect the weight of work put into amelioration of "sick" nurseries may be questioned. But at the outset the Commission's stake in old agricultural soil type nurseries was considerable, and also it was thought possible that new nurseries might decline quickly with intensive cropping. Though the performance of new nurseries under the best manurial practice has not given rise to much anxiety, it is undoubtedly valuable to have so much knowledge on the biological factors, which we can apply in cases where for one reason or another it is not desirable to shift to new ground.

From the scientific aspect, the NNC programme is an excellent example of the value of calling for new thought in an apparently intractable situation. Crowther was free from the assumptions which had grown up around the subject of nutrition of conifers, and immediately posed a fresh set of questions. The Rothamsted part of the programme is an interesting and profitable example also of research designed and directed by one institution and executed by another. Few collaborations of this sort can have had happier or more fruitful results.

The later stages of Benzian's investigations may be said to have explored the limits of nutrition in the production of bare rooted nursery stock by conventional methods in the uncontrolled nursery environment. Seasonal variations of growth (especially of seedlings) though not easily related to climatic factors are quite large compared with nutritional effects, and it is plain that if anyone erects a further target—say the plantable one-year-old seedling—some degree of environmental control of the growing season will be necessary.

One valuable by-product of the NNC programme was the introduction to forest research by Crowther of the factorial designs developed by Yates at Rothamsted. The programme also provided an intensive and stimulating course in experimentation for the younger Commission research workers collaborating with Rothamsted.

REFERENCES

BENZIAN, B. (1965). *Experiments on nutrition problems in forest nurseries.* 2 vols. Bull. For. Commn, Lond. 37.

BENZIAN, B. (1966a). Manuring young conifers: Experiments in some English nurseries. *Proc. Fertil. Soc. 94*, 5–37.

BENZIAN, B. (1966b). Risk of damage from certain fertiliser salts to transplants of Norway spruce and the use of slow-release fertilisers. *Suppl. Forestry*, 65–69.

BENZIAN, B. (1967). Test on three nitrogen fertilisers —'Nitro-chalk', formalised casein and isobutylidene diurea—applied to Sitka spruce (*Picea sitchensis*) seedlings in two English nurseries. *Proc. 5th Colloqu. int. Potash Inst.*, Finland, 171–175.

BENZIAN, B., and BOLTON, J. (1966). Calcium as a plant nutrient for Sitka spruce. *Rep. Rothamst. exp. Sta. Harpenden* 1965, 62.

BENZIAN, B., BOLTON, J., and MATTINGLY, G. E. G. (1969). Soluble and slow-release PK fertilisers for seedlings and transplants of *Picea sitchensis*. *Pl. Soil 31*, 238–256.

BENZIAN, B., and FREEMAN, S. C. R. (1967). Effect of "late-season" N and K top-dressings applied to conifer seedlings and transplants, on nutrient concentrations in the whole plant and on growth after transplanting. *Rep. Forest Res., Lond.*, 135–140.

BENZIAN, B., and FREEMAN, S. C. R. (1970). Nitrogen concentrations in conifer transplants and subsequent growth in the forest. *Rep. Forest Res., Lond.*, 168–170.

BENZIAN, B., FREEMAN, S. C. R., and MITCHELL, J. D. D. (1971). Isobutylidene diurea and other nitrogen fertilizers for seedlings and transplants of *Picea sitchensis* in two English forest nurseries. *Pl. Soil. 35*, 517–532.

BRIND, JANET E. (1965). Some studies on the effect of partial sterilization on the soil micropopulation. In *Benzian* (1965), 206–211.

GOODEY, J. B. (1965). The relationships between the nematode *Hoplolaimus uniformis* and Sitka spruce. In *Benzian* (1965), 210–211.

GRIFFIN, D. M. (1955–1957). Fungal damage to roots of Sitka spruce seedlings in forest nurseries. *Rep. Forest Res., Lond.* 1954, 52–53; 1955, 75–76; 1956, 86–87.

GRIFFIN, D. M. (1965). A study of damping-off, root damage and related phenomena in coniferous seedlings in British forest nurseries. In *Benzian* (1965), 212–227.

LEVISOHN, I. (1952–1961). Researches in soil mycology. *Rep. Forest Res., Lond.* 1951, 123–126; 1952, 119–120; 1953, 102–103; 1954, 50–52; 1955, 76–78; 1956, 83–86; 1957, 96–98; 1959, 91–94; 1960, 84–87.

LEVISOHN, I. (1965). Mycorrhizal investigations. In *Benzian* (1965), 228–235.

RAM REDDY, M. A. (1963). A study of the effects of disease control measures on the soil microflora of Sitka spruce (*Picea sitchensis*) in forest nurseries. *Rep. Forest Res., Lond.* 1962, 89–90.

RAM REDDY, M. A., SALT, G. A., and LAST, F. T. (1964). Growth of *Picea sitchensis* in old forest nurseries. *Ann. appl. Biol.* 54, 397–414.

SALT, G. A. (1964–1970). Pathology experiments on Sitka spruce seedlings. *Rep. Forest Res., Lond.* 1963, 83–87; 1964, 89–95; 1965, 97–102; 1966, 104–108; 1967, 141–146; 1968, 156; 1969, 147; 1970, 174–175.

WARCUP, J. H. (1951). Effect of partial sterilization by steam or formalin on the fungus flora of an old forest nursery soil. *Trans. Br. mycol. Soc.* 34, 519–532.

WARCUP, J. H. (1952). Effect of partial sterilization by steam or formalin on damping-off of Sitka spruce. *Trans. Br. mycol. Soc.* 35, 248–262.

Chapter 14

OTHER NURSERY INVESTIGATIONS

Nutrition was undoubtedly the most important nursery subject in the post-war period and attracted the greatest share of the work. But a number of other lines were followed with large effects on nursery economy. Most of the work in the period accepted the traditional pattern of nursery working; i.e. the production of bare-rooted stock by sowing in the open bed and transplanting, though attention was given to undercut stock as an alternative to the transplant. But by the end of the period certain lines of work appeared which bore little resemblance to traditional nursery practice, and introduced some degree of environmental control.

The order in which the topics will be discussed is arbitrary, but it appears convenient to deal first with the most important subject—weed control.

Weed Control

It will be recalled that the blow-lamp had been used to some extent before the war as a pre-emergent treatment on seedbeds: and experimental treatments with sulphuric acid and sodium chlorate had shown some promise. The first big post-war success in controlling weeds in seedbeds was in fact steam sterilisation: a useful by-product of what was primarily a soil ameliorative treatment.

Certain mineral oils had been used for weed control in agriculture, especially on carrot seedlings, and in 1948 Holmes started work on the oils; the first application being as post-emergent treatments on conifer seedbeds, looking for some degree of selectivity between conifer seedlings and the common weeds. But by 1950 it was clear that the greatest promise lay in the use of oils as pre-emergent treatments; that is used unselectively to control weeds germinating before the crop. After successful large-scale trials in 1951 the type known as vaporising oil became a standard treatment for seedbeds in all Commission nurseries with high weed populations. Vaporising oil proved safer and more efficient than flame gun treatment, and lowered weeding costs in the seedbed by as much as 80 per cent, besides reducing seedling losses by postponing the first hand-weeding till seedlings were less likely to be removed with the weeds. This treatment alone was capable of giving a net saving of some £50 per acre of seedbed (Holmes and Ivens, 1952). Plate 34.

The white spirits showed promise of being sufficiently selective for use as pre-emergent treatments in conifer seedbeds, but the timing of the operation required a good deal of care, and this treatment never assumed the importance of the pre-emergence application of oils. Search was continued for truly selective herbicides for use in seedbeds, and it was Aldhous' practice to screen virtually all new herbicides with any prospects from the rather bewildering range offered by the chemical industry in the fifties and sixties. This was aiming at a receding target, since sowings were all the time shifting to (initially at least) weed-free nurseries, but the work was cheap and a truly selective herbicide would always have been worth money. By the close of the period such a substance had not been found.

The other main target in the nursery was the transplant lines, and here there was a chance of adapting horticultural machinery (light mechanical hoes or tillers) for forest nursery crops, since it was possible to control spacing to suit machines. Though a good deal of attention was given to this by nursery managers and mechanical engineers, the position was transformed by the introduction of simazine. This general, residual, herbicide has the valuable property of being held close to the surface of the soil, and hence selects between germinating weeds and deeper rooted plants. It was first tested in forest nurseries by Aldhous in 1958, and came into general use in the early sixties after intensive work on rates, mode of application, and possible hazards. A few species appeared particularly susceptible to simazine, notably poplar and ash, but on the whole this herbicide gave little trouble, and it was difficult to detect any build up of harmful residues at dosages anywhere near the standard effective rates.

The herbicide paraquat appeared in nursery programmes a year or two later than simazine, and proved a useful alternative to vaporising oil as a pre-emergent treatment to seedbeds. The desiccants paraquat and diquat assumed greater importance in the forest, but have been of considerable use in the nursery for the control of weeds in paths, surrounds and on fallow ground.

One interesting enquiry by Aldhous might have been undertaken profitably in the twenties. This was a comparison of frequencies of hand-weeding in the seedbed at intervals in the range of 2 to 8 weeks or more. It was shown that the *total* weeding time for the season was very much less at the shorter intervals (round about three weeks) than at the longer intervals; an advantage which was apparent well before the weeds had developed to the detriment of the crop.

It is difficult to put a cash value on the work on herbicides, because of the change towards cleaner nursery sites which was proceeding concurrently. In all established nurseries which continued to be cropped the benefits were very large indeed, and the appearance of these nurseries at the end of the period is a dramatic improvement on that of the early fifties. As it happened, chemical methods in the nurseries have owed little to true selectivity; the important herbicides have had to fit into the husbandry of the system.

Storage and Handling of Plants

The traditional method of storage of lifted seedlings or plants awaiting lining out or planting in the forest was by heeling-in closely packed rows of plants in shallow trenches (Scottish: sheughing). Carefully done, the method was quite satisfactory, but it often took up a good deal of nursery space, and the stocks might be frozen in or covered with snow when wanted somewhere else. The heavy production from some of the early heathland nurseries encouraged enquiry into more convenient storage methods; and a purely domestic Research incentive was the need to store experimental seedlings off the ground to permit early lifting and cultivation of the unit plots. It was found quite practicable to store such seedlings in boxes in airy sheds using peat as a packing medium; this, however, was far too fussy a method for large-scale use.

It was known that American nurserymen, accustomed to severe winters, had developed the practice of lifting in the fall, and grading and packing under cover; storing the wrapped plants in controlled conditions. We might have copied some of these methods earlier than we did, but for our normally open winters and the practicability of doing a good deal of the work outside.

In experiments on the storage of seedlings, Faulkner used an old salmon store in 1954—a primitive cold chamber. The refrigerated seed stores at Alice Holt were first used for the experimental storage of plants in 1959, after a few seasons work with polythene film for packing and short-term storage. Studies on exposure to drying had shown that survival rates of conifer seedlings were fairly closely related to moisture loss—as would be expected. Polythene bags equalled the best conventional forms of packing, with great saving of time. Curiously enough the use of latex dips never offered much promise with conifers. Polythene established itself as a packing material, and for short-term storage (a few weeks), almost immediately; long-term storage (months) was the next interest. It was soon shown that most conifer seedlings or plants, if packed without moisture on the foliage-tolerated storage for several months under normal winter temperatures. The cold chamber operating at about 2°C was the obvious refinement, and once the responses of plants had been worked out to season of lifting, length of storage, and season of lining-out or planting, certain added advantages appeared which justified the use of refrigerated stores. The flexibility of the system

was markedly increased. Seedlings could be held dormant till late in the spring when weather or other conditions had delayed the lining-out programme. Also, when there were large surpluses of seedlings, they might be "banked" by lifting in spring and storing dormant till July, when lining-out gave good survival rates but little growth in the remainder of the season. Plants held dormant in cold chambers extended the safe planting season, especially under Scottish conditions.

Physical Conditions of the Seedbed

A number of investigations can be lumped under this title, for want of a better. These were matters of technique, all of which had received some attention before the war, but which appeared to require further study in the light of changed conditions, notably the improvement in growth of seedlings brought about by nutritional studies.

(1) *Seedbed Covering Materials.* The dismissal of calcareous covering materials used in certain nurseries has been mentioned under nutrition in Chapter 13. The main development was the shift from sands to grits, which change had started in Scotland before the war. This was largely a precaution against windblow. Some further work on depth of cover and particle size was required. St Austell quartz grit (a residue of kaolin mining) became a standard medium in Research and many other southern nurseries, and elsewhere other suitable covering materials were identified. Very dark covering materials were shown to produce dangerously high surface temperatures on occasion. (Plate 39.)

(2) *Irrigation.* Work on irrigation was carried on over almost the whole period. It was very much a Kennington speciality, though a few collaborative trials were carried out in production nurseries in England. In Scotland the few experiments carried out were discouraging. The work developed from small-scale watering experiments to the use of modern agricultural sprinkler systems. In the later stages watering regimes were controlled by measured (tensiometer) soil deficits, or deficits calculated by Penman's formula. Date of sowing was commonly included as a treatment, and some work was done on the application of nitrogen through the irrigation system. The result can be summed up very briefly. Under southern English conditions, with moderate-to-low rainfall and high evaporation rates, irrigation almost invariably increased the yield of seedlings. More often than not it also increased the height of seedlings, but in unusually wet summers growth responses might be small or absent. The benefits of watering were usually least in early sown treatments and this was an important point in considering irrigation as a practice in production nurseries. It did not appear to be the critical factor in obtaining

well stocked beds of usable seedlings frequently enough to invest in the apparatus and water supplies. However, as with certain other lines of work (such as partial sterilisation) studies on irrigation provided experience which might be useful in permanent, specialised nurseries in the drier parts of the country.

(3) *Date of Sowing.* This was an old topic, and pre-war work had pointed towards early sowing dates. Better seedbed covering materials (some of the old work had in fact been done with soil covers), and seed in better condition through improvement in storage methods, were factors which might be expected to influence fresh enquiries on dates of sowing. In fact post-war work under improved seedbed conditions showed fewer anomalies, and though there were the seasonal and regional differences which one would expect, the great weight of evidence favoured sowing dates in late March and early April; the higher yield of seedlings being the most obvious effect, but growth also responding to early sowing. A few experiments were done on *season* of sowing, comparing autumn with spring sowings, but though one or two species appeared to tolerate autumn sowing, the results were not encouraging.

Experimental comparisons of sowing dates were maintained in Research nurseries well after the necessary evidence for practice had been obtained, because they provided a useful reference for the behaviour of seedlings sown at different dates in other experiments; and, with irrigation experiments, gave some evidence on the characteristics of the season for growth.

(4) *Protection of Seedbeds.* Lath screens were commonly used before the war both for winter frost protection of seedlings and for shading against sun-scorch in summer. Experimental evidence on neither practice had been consistent. No research was done after the war on frost protection; lift had always been the most damaging effect of frost, and it was observed that this simply did not occur with well grown seedlings, following the advances in nutrition. Shading on the other hand attracted a few experiments in the late forties and early fifties. Slight benefits were recorded on occasion, but usually with late sowings; it appeared clear that the practice was not worth while with early sown seedbeds.

Density of Sowing

Here again there was a body of pre-war experimental work on which practice had been based, but improvements in the germination and survival rates in the seedbed necessitated some reconsideration of sowing densities; these were usually too high.

The nursery side of the work tied in with the Seed Laboratory's studies on germination testing and prescription of sowing density. The special aspect of sowing density for beds destined to be undercut

was studied in Scotland; this included comparisons of drill (or band) sowing versus the conventional broadcast. Though drill sowing always appeared the logical method for mechanised nurseries, at no time did the Commission succeed in obtaining yields of seedlings equal to those from broadcast sowings in large-scale working. This was most obvious with the smaller seeded species such as Sitka spruce, and one probable cause was the difficulty of covering drills with the same accuracy as broadcast beds.

Grading

It was mentioned in Chapter 13 that definition of quality of stock always presented difficulties. A great deal of work had been done on grading before the war, but this had undoubtedly suffered from the use of subjective characters in comparisons; e.g. the term "cull" did not mean the same thing to different people. Post-war work certainly did not lead quickly to a set of objective standards, but being based on better statistical theory it did at least deal with quantitative characters expressed in terms of the population.

The importance of collar diameter as against height of seedlings as a survival factor was shown by Holmes in 1949. It was later shown by A. J. Rutter (1955) to bear a better relation than height to dry weight in young plants. Crowther had always appreciated that dry weight would have been the logical assessment of experimental seedlings, but there were practical difficulties in using weight.

Probably the most important post-war work (other than nutritional) on quality was associated with the investigations on root pruning and undercutting, mentioned below.

Undercut Stock

Before the war, root pruning, or its crude form wrenching, had been practised occasionally to hold back surplus stocks from growing too large. Little had been done on undercutting as an alternative method to transplanting for the production of planting stock, though a good deal of attention had been paid to seedlings. Post-war work started in the early fifties, examining depths and seasons of undercutting, density of sowing etc. for a considerable range of species. Undercutting was performed with a tractor drawn sledge-mounted blade, and the aim was usually to produce a two-year-old plant undercut between the growing seasons, for comparison with two-year-old undisturbed seedlings and conventional "one plus one" transplant stock. Once a satisfactory machine had been developed the method gave no difficulties. (Plate 37.)

Measurements of the three types of stock placed the undercut plants in an intermediate position for a number of characters, notably root/shoot ratio; and it was expected that this would be reflected in the survival rates of the stocks when experimentally planted in the forest. Surprisingly, this was not the case with any species except the difficult Corsican pine, which behaved according to expectation, transplants being better than undercut stock and undercut stock than seedlings. For the rest; a quite considerable investigation had failed to show why we should not plant two-year-old undisturbed seedlings, let alone undercut stock (Atterson, 1964). This enquiry had very little effect on practice, since it continued to be believed that, under the greater hazards of large-scale planting, the transplant would have a worthwhile advantage. A few nurseries, however, notably Ledmore in Perthshire and Taironen in South Wales, produced undercut stock on considerable scale with success.

Intensive Nursery Methods

It was always appreciated that the conventional open seedbed was at the mercy of the elements, but environmental control seemed an unrealistic target in view of the scale of our nursery work, and the relative cheapness of the system. However, some tentative approaches were made by Edwards and Faulkner in Scotland in the late forties and early fifties. These aimed at the production of large one-year seedlings, or two or more crops of usable seedlings in one season, in frames with soil heating. Latterly Dutch lights illuminated and slightly heated by ordinary electric bulbs were tried, with considerable success. The work was useful enough in giving an idea of the light factor in the growth of seedlings. Had there been an obvious economic objective to pursue at this time no doubt systems based on the application of a little soil heat could have been worked out. These would, it is certain, have guaranteed the production of large one-year seedlings of Sitka spruce, whereas (especially in the north) it has never been possible to be sure of 100 per cent production of usables, however good the nutritional and other techniques.

Interest in unconventional methods revived in the late sixties. By this time the Genetics Section had removed the germination phase of their valuable seed from the open ground to the glasshouse, and were lining out seedlings (using a miniaturised lining-out board) a few weeks old, out of germinating dishes. This development, whilst of much interest, did not appear capable of scaling up. It was by now obvious that there was no theoretical bar to the removal of seedling production from the nursery altogether, possibly to some continuous system involving cold storage. However, the first large-scale work on plants raised under controlled environment had more prosaic objects, and was concerned with a particular type of container, the small diameter plastic tube which had been tried extensively in Canada. A primary incentive was to establish Lodgepole pine of coastal provenance in such a way that the plants did not become subject to wind-sway at an early age. It was thought that very young seedlings might develop a better root system than conventional transplants. The main disadvantages of the older types of container are of course transport and handling costs, besides the large space taken up at the stances. These tubes of 16 mm ($\frac{1}{2}$in) diameter however take up little growing space, and offer one real prospect for the transference of stock grown under controlled environment to the forest; that is to such sites as are suited to very small seedlings. Because the limited planting season can be doubled the use of these "tubelings" enables a finite labour force to achieve twice the programme. An interesting development in this programme (which continues with a range of species) has been the use of a polythene sausage-shaped greenhouse for temperature, light, and ventilation control. This erection somewhat surprisingly withstood the gale of January 1968 associated with Low Queenie. (Plate 36.)

Another imported idea is the Nisula plant roll, of Finnish origin. Seedlings are placed on fertilised peat on long bands of polythene sheet, which are then rolled up like a Swiss roll. The rolls are tended for a season, producing something analagous to 1 plus 1 transplant stock, but with an entire and superior root system. Early indications are very promising, and this method appears very flexible, since it can make use of varying degrees of controlled environment in both the seedling and roll stage.

Under British conditions, all methods using environmental control to any degree are up against the fact that conventional nursery methods are cheap and fairly reliable; and bare rooted stock produced by these methods is very successful. The seedbed is undoubtedly the least reliable part of the system, and to replace it for seed of high genetic value seems an obvious aim. Whether the seedbed (or the lines) can be replaced for large-scale production remains doubtful in view of the capital cost of the installations. It seems probable that the key to any successful alternative system would be the use of continuous production and storage.

Pests and Diseases

Apart from the work on soil pathogens (mentioned in Chapter 13) the nursery did not attract or require much attention from pathologists or entomologists during this period. Probably due to changes in soil types of nurseries and the abandonment of leys and green crops, the chafers, which were common pests in England in the pre-war period, ceased to have any importance. The cut-worms (Noctuid moth larvae) remained and occasionally did serious damage to seedbeds. They were amongst the many pests which were successfully controlled by insecticides of the organo-chlorine group, now very much under attack, but probably an insignificant hazard in forest nurseries.

A new source of damage discovered in the post-war period was the attack on very young seedlings by springtails (*Collembola* sp); these minute creatures might often evade detection. Where identified, they were readily controlled by malathion.

Bird damage to seedbeds (often very serious) attracted work on retardant chemicals, usually applied as seed dressings; but no wholly reliable preparation was found. The appearance of polypropylene netting provided the obvious answer.

Miscellaneous

The Dunemann bed—a raised board-sided seedbed composed of spruce litter—attracted some attention in the late fifties, having given good results on certain private estates. It produced very good one-year seedlings of many conifer species, but required a great deal of attention to maintain optimal moisture conditions. It was perhaps a useful technique for private estates without suitable seedbed soils, but otherwise too fussy altogether. It is possible that the Dunemann bed would have been regarded as an important innovation before the war.

Pre-war work on vegetative propagation of conifers by cutting was followed up in silviculture for a year or two after the war, but the main advances were made later by the Geneticist in developing techniques for the breeding programme.

One curious side-line was the investigation in the mid-fifties on the growth inhibitor, maleic hydrazide, to hold back surplus stocks, as an alternative method to the traditional wrenching. This was not without promise, but the condition of plants after treatment was not satisfactory. Developments in storage methods and better stocktaking overtook this line of enquiry.

Machinery

Developments in mechanisation of nursery operations were not the direct concern of the Research Branch, but were shared between the Headquarters Machinery Research Officer, the Directorate Mechanical Engineers and numerous field officers. The main line was the adaptation of a number of appliances to a three-point linkage for the Ferguson agricultural tractor. Initial cultivations, preparation of beds, sowing, covering, are all satisfactory carried out mechanically, and lifting of stock can be facilitated. Although various approaches have been made, mechanisation has until very recently stopped short of full transplanting, which is still generally done by ploughing and man-handled lining-out boards.

It would be difficult to apportion accurate shares in the improvement of the general nursery economy between the various technical advances, or even between nursery technique, nutrition, and seed. Nor should one forget improved nursery management. However, the period was one of steady advance in nursery methods punctuated by several outstanding research achievements. Aldhous (1967), in reviewing developments in nursery technique over the years 1949–66, gives the following table which speaks for itself.

TABLE 1

AREA OF FOREST NURSERY, PLANTS USED AND COSTS OF NURSERY
WORK, FOR THE FORESTRY COMMISSION 1949–1965

Year	Total nursery area Acres	Number of plants used Millions	Actual expenditure, excluding overheads £000	Cost in terms of the value of the £ in 1965
1949	2,233	83	£518	£865
1954	2,129	122	£517	£699
1959	2,100	105	£542	£635
1964	1,768	109	£402	£420
1965	1,712	99	£371	£371

REFERENCES

ALDHOUS, J. R. (1967). *Review of research and development in forest nursery techniques in Great Britain, 1949–1966.* Res. Dev. Pap. For. Commn, Lond. 46.

ALDHOUS, J. R. (1972). *Nursery practice.* Bull. For. Commn, Lond. 43.

ATTERSON, J. (1964). Survival and growth of undercut seedlings in the nursery and forest. *Rep. Forest Res., Lond.* 1963, 147–156.

HOLMES, G. D., and IVENS, G. W. (1952). *Chemical control of weeds in forest nursery seedbeds.* Forest Rec., Lond. 13.

LOW, A. J. (1971). Tubed seedling research and development in Britain. *Forestry* 44 (1), 27–41.

RUTTER, A. J. (1955). The relation between dry weight increase and linear measures of growth in young conifers. *Forestry* 28 (2), 125–135.

Chapter 15

SPECIES

The general lines of work bearing on choice of species have been mentioned in Part I, Chapter 6. In the post-war period the choice of species hardened somewhat, spruces and pines accounting for over 80 per cent of the plantations at the end of the period as against about 75 per cent at the beginning, with some decline in the planting of larches. The only dramatic change in the status of an individual species was the rise of Lodgepole pine from insignificance to second behind Sitka spruce at the end of the period. Planting of hardwoods declined to a very low figure as acquisitions of land suited to broadleaved species virtually ceased. This superficial comment hides a good deal of adjustment in species boundaries, but with the important exception mentioned, it is the case that there was very little movement in species suited to the *afforestation* of major land types during the period.

Research work on choice of species is of course not purely a silvicultural preserve; nearly all research activities have a bearing on it. But it appears convenient to review the primarily silvicultural work together. During the period there was an obvious trend towards breaking up the work into projects concerned with special circumstances, and by the end of it many species seemed more likely to be of importance in arboriculture rather than economic forestry.

Species in Afforestation

Trials of species were associated with virtually all investigations into establishment on special sites. Frequently the species comparison was subsidiary to the cultural treatments, e.g. the experiment might be split for two or more species. On very dubious sites (as in many of the pilot trials in the north of Scotland) species under trial were planted in a matrix of a safe choice; Sitka spruce or Lodgepole pine. At this stage there was little expectation of extending the plantable range by new species; there was, however, a chance that something might be found to perform as well as or better than the accepted choice under adverse conditions. For example, some attention was paid to species of apparently low nutritional requirements (judged on behaviour in their native habitat) on the poorest peats. *Tsuga mertensiana* and *Chamaecyparis nootkatensis* were tried in this role. A few mountain species were also tried at high elevations. On the whole this approach has simply confirmed the position of Sitka spruce and *Pinus contorta*, and certainly nothing has appeared which is likely to make any mark on the national scale in afforestation. *Abies procera* has

performed quite well on mineral soils at high elevations in Wales (especially) but is expensive to raise and slow to establish. In the climate of the south-western coastal regions it has become clear that *Pinus radiata* has a place. Some attention has been given to the interesting bi-generic hybrid *Cupressocyparis leylandii*, and this has been established in trials over a wide range of soils and climatic types. That it will form productive plantations on many sites is not in doubt, but it is not yet clear whether there is any site on which it would be the best choice on silvicultural grounds, and there is also the economic question whether a somewhat "special" timber is to be welcomed.

On the whole, in the field of coniferous afforestation more has been gained in the post-war period by work at the provenance level than by the study of new species.

Conifers in Regeneration

With the exception of Norway spruce, all the chief afforestation species are intolerant trees belonging to the early ecological succession in their native habitats. The tolerant late succession species have (understandably) proved difficult to establish in bare land planting, and have played little part in afforestation. However, there has been interest in the north-west American tolerants, especially *Abies grandis*, Western hemlock and Western red cedar, from their success in arboreta and in pioneer plantations, and these trees (amongst others) have been widely planted in forest plots; and they have appeared on a minor but significant scale in the conversion of broadleaved scrub and coppice and in underplanting of larches etc. Western hemlock and *Abies grandis* appeared likely to prove very high yielders, and this was confirmed by the publication of provisional yield tables by Evans and Christie (1957) and Christie and Lewis (1961). In addition to such mensurational studies, Research has been interested in finding the useful range and requirements of the most promising minor species, and has planted them over a range of sites. In particular, they have appeared in many of the regeneration experiments. It has been shown that the climatic range of *Abies grandis* and Western hemlock is wide, and neither has the high moisture requirement of Sitka spruce. Western hemlock also seems to tolerate less fertile conditions than Sitka spruce, and has made a fair showing on heathland sites. These characteristics would be expected from the behaviour of the trees in their natural range (Wood, 1955).

In the late sixties, Research undertook an evalu-

Plate 27. Control of Pine weevil, *Hylobius abietis*, during the pre-war period with the use of log traps gave a fair measure of protection. Ch. 8. D3206.

Plate 28. Trial of Macedonian pine, *Pinus peuce*, at Beddgelert Forest, Caernarvonshire. The choice of species is the most important of all management decisions. Before the 1939–45 war a number of species were considered, but in the post-war period the choice of species hardened on a very limited number. Ch. 6. D2337.

Plate 29. Old Scots pine hedgerow on the Brandon/Bury St Edmunds road, Thetford Forest, East Anglia. It was noted, very early on, that despite past clipping the progeny of the old Breckland hedgerow Scots pine grew well not only on its home ground at Thetford but also under very different conditions in Scotland. Ch. 6. C3239.

Plate 30. Classifying conifer seeds by weight. Apparatus for classifying conifer seeds into 1 milligram classes. Containers, with trip levers above at varying heights, run along on a trolley. Seeds placed on a glass fibre balance (modified to achieve oil damping), are tipped into a given container when the height of the balance pan coincides with the appropriate lever. Ch. 12. D5797.

Plate 31. Germination testing of seed at Alice Holt Lodge, Hampshire. The Seed Laboratory set up in the early 1950s initially tested only the seeds to be sown in research experiments, but in 1951 was sufficiently well established to be licensed as a Private Seed Testing Laboratory. Since then all seed for use or sale by the Forestry Commission has been tested there. Ch. 12. D6286.

Plate 32. Steam sterilisation of nursery beds at Ampthill Forest, Bedfordshire, 1949. The effect of steaming on the growth of seedlings of various species is demonstrated (e.g. the Japanese larch on the left of the standing figure has been treated, but the same species to the right has not). Ch. 13. C936.

Plate 33. Glasshouse progeny trials. Heated propagation tray containing Sitka spruce progenies in separate PVC seed pans, photographed six weeks after sowing. Sitka spruce has had a more constant seed source than any other American species, with the Queen Charlotte Islands having provided the main supply and Coastal Washington a much smaller proportion. The potential gain in production in taking seed from Washington was considerable, and the main object of subsequent provenance work in Sitka was to assess the risk from more southerly seed sources. Ch. 22. C4515.

Plate 34. Nursery spraying of herbicidal oil at Kennington Nursery, Oxford, in 1950. The herbicide is put in the small glass container mounted on a traveller which traverses the plot in two directions at right angles. Air pressure is supplied from a fine pressure bottle and the herbicide emerges from a special jet on the traveller in a fine spray. Ch. 14. C76.

Plate 35. Soil sterilisation with steam, using Hoddesdon equipment at Ampthill Nursery, Bedfordshire. This was one of the principal ameliorative treatments applied to "Sitka sick" nurseries. Although steaming was highly successful, it was costly and inconvenient and abandoned in favour of formalin by 1953. Ch. 13. B347.

Plate 36. "Tubling" production of seedlings. Small diameter plastic tubes have been tried out extensively in Canada and were developed in the late 1960s. The tubes, of 16 mm diameter, take up little growing space and offer one real prospect for the transference of stock grown under controlled environment to the forest. They have proved especially effective on peat planting, where smaller plants are acceptable, and the production period has been reduced from two years to under six months. Ch. 14. A3720.

Plate 37. Undercutting one-year seedbed of Lodgepole pine sown thinly for production of 1-undercut-1 planting stock at Glenfinart Nursery, Argyll. Work started in the early 1950s examining the depth and seasons of under-cutting, density of sowing etc., for a considerable range of species. Once a tractor-drawn, sledge-mounted blade had been developed the method gave no difficulties. Ch. 14. C3997.

Plate 39. Hand covering seedbeds with riddle at Alice Holt Lodge, Hampshire, in 1953. Investigation into seedbed covering materials resulted in the dismissal of calcareous covering material, the shift from sands to grits as a precaution against windblow, and avoidance of very dark covering materials which produced dangerously high surface temperatures. St Austell quartz grit (a residue of kaolin mining) became a standard medium in Research and many other southern nurseries. Ch. 14. A1497.

Plate 38. Storage and transport of nursery stocks. Studies on exposure to drying showed that survival rates of conifer plants are closely related to moisture loss. The use of polythene bags was developed for both short-term and long-term storage in the late 1950s. Ch. 14. B3741.

Plate 40. Poplar variety trials at Quantock Forest, Somerset, planted in 1950. *Populus trichocarpa* in the foreground. Poplars, as with elms, are normally grown outside of the forest on land of agricultural quality. Much of the experimental work involved testing clones for susceptibility to canker. Ch. 15. B1282.

Plate 41. Southern beech, *Nothofagus obliqua*, at Stonedown Wood, Cranborne Chase, Wiltshire, 14 years old. This species and *N. procera* are late introductions to Britain from Chile. They have shown considerable potential as broadleaved trees growing at "coniferous rates" on sites outside the normal productive range of broadleaved species in Britain. A wide scatter of plots were established throughout the country, and interest has increased with the emphasis on the use of broadleaves for amenity. Ch. 15. B5862.

Plate 42. Exposure flags used for advanced estimates of exposure on questionably plantable sites in Argyll. Ch. 16. B4416.

Plate 43. Afforestation of opencast spoil in South Wales. A variety of industrial sites of importance in local amenity were investigated in the early 1950s. There appeared to be little or no advantage in waiting for weathering or ecological succession to advance before planting. Ch. 16. C4628.

Plate 44. Aerial application of fertilisers to checked plantations at Wilsey Down, Kernow Forest, Cornwall, in 1959. A fixed-wing Tiger Moth aeroplane in operation, distributing 3 cwt of triple-superphosphate per acre. This was the first British experience of aerial fertilisation of forest crops. Ch. 17. A2164.

Plate 45. Land ploughed with the single-furrow "RLR" plough at Broxa, Langdale Forest, Yorkshire. A five-year-old plantation of Sitka spruce and Scots pine treated with basal Gafsa phosphate at the time of planting. Ch. 17. B2899.

Plate 47. Treatment of checked Sitka spruce at Wilsey Down, Kernow Forest, Cornwall. 6 cwt per acre of triple-superphosphate were applied five years previously to 20-year-old checked Sitka spruce by a tractor-mounted spinner. The great success of this exercise led to widescale treatment of checked plantations and to the increasing use of phosphates at the time of planting. Ch. 17. C3827.

Plate 46. Checked Sitka spruce, 28 years old, at Wilsey Down, Kernow Forest, Cornwall. Despite the stunted growth, repeated dieback and lichen-covered trees, these plantations were shown to respond to phosphate manuring. Ch. 17. C3831.

Plate 48. Mechanical scrub clearance with the V-blade scrub cutter mounted on 120 hp Allis Chalmers tractor at Queen Elizabeth Forest, Hampshire. The cutter is shaped like a snow-plough with a sharp horizontal blade, which cuts the scrub stems at ground level. This was a most adaptable machine developed by A. D. Miller and D. Hampton. It was capable of cutting stems up to 9 inch diameter and would cut its way through scrub or coppice with comparatively little of the uprooting which was so objectionable, causing much local puddling on heavy soils. Ch. 19. B3524.

Plate 49. Derelict woodland rehabilitation experiment at Flaxley, Forest of Dean, Gloucestershire, in 1956. Treatments include clear-felling, mechanical clearance of strips and replanting with various combinations of species. Ch. 19. B5412.

Plate 51. Underplanting of larch at Dyfi Forest, Merioneth. In the extreme light demanders, such as larch, underplanting at a suitable stage with a shade tolerant species was a well established practice before the war, and a number of sample plots were established to study the productivity of such two-storied crops. Ch. 21. B5294.

Plate 50. High pruning of Norway spruce at Drummond Hill Forest, Perthshire. Economic evaluation of high pruning has long been considered, and the first experiments were carried out by Guillebaud in the early 1930s. Ch. 20. D5055.

Plate 53. Oak spacing experiment showing 3 ft × 2 ft (1 × 0.66 m approx.) spacing plot 18 years after planting in 1936. A few spacing experiments had been established in the early 1920s, but the main effort dates from 1935 when James Macdonald drew up a plan for a standard type of spacing comparison. Ch. 21. C2340.

Plate 52. Aerodynamic studies at Redesdale Forest, Northumberland. Gales gusting to 60 miles per hour occur somewhere in Britain every year. The behaviour of crops in such normal gales is of particular research interest. The forces on the individual tree were investigated by measuring the wind-loading dynamometers on guide stands. Ch. 21. C4155.

Plate 55. The development of female flowers on Sitka spruce. Studies in tree physiology were begun by Prof. P. F. Wareing of Manchester University (and later Aberystwyth) in 1956. He investigated the factors which govern the onset of maturity (flowering and seedbearing) and those which induce flowering in trees which were physiologically ready for it. Ch. 22. D6375.

Plate 54. Cuthbertson double-furrow drainage plough at South Kintyre Forest, Argyll. The Cuthbertson ploughs were introduced in 1945 and were one of the key developments in moist upland afforestation. Single-throw ploughs were designed for true drainage; double-throw machines as above, provided turf from shallow furrows to give drainage, a raised planting position, physical suppresion of vegetation and manuring from the grass rotting beneath the furrow. Ch. 18. C4053.

ation of the most prominent of the minor species (Western hemlock, Western red cedar, *Abies grandis* and *Abies procera*) taking into account silvicultural, pathological and economic factors. This was something of a departure since species choice had usually evolved, whilst this was an effort to quantify the available information in order to allow policy decisions whether to plant these species on a forest scale or no.

This study, conducted by Aldhous and Low, involved the study of about 500 sites on which one or other of the minor species might be compared with the normal choice for that site. All these trees had already been studied in the joint programme on timber properties at Princes Risborough, and Pathology had also made important studies on the incidence of *Fomes* rot, and (in the *Abies* sp) of stem crack associated with drought. At the time of writing, the final report has not been published, but findings have been summarised by Aldhous and Low (1970). It is of interest that the criterion has been reduced to profitability of production. Forty years ago the important argument for the use of these species in British forestry appeared to Sir John Stirling Maxwell (1932) to be the diversification of forest crops on general ecological grounds.

Broadleaved Species in the Forest

The general decline in the economics of hardwood forestry has discouraged work on broadleaved trees. They have, however, remained of research interest for special circumstances. One of these, timber production outside of the forest, has attracted a great deal of work on poplars and elms, which is dealt with later.

Apart from the ancient introductions, sycamore and Sweet chestnut, no exotic broadleaved tree has made any great impact on British woodlands. The Commission's hardwood policy was based mainly on beech and oak, and there was no incentive to look for alternative species on the best beech and oak sites. As mentioned in Part I, Chapter 4, some attention was given to broadleaved species as pioneers for beech woodland on the Chalk. The one outstanding success was *Alnus cordata*, and this is a tree which should be used more frequently wherever fast broadleaved cover is required, as for instance on industrial sites.

Outside of the main hardwood areas, one of the early interests in broadleaved species was to provide nurse or pioneer crops for the principal conifers. For the original object, this was an abortive line, but it did throw up one or two species whose performance was of interest. *Alnus rubra*, an associate of the successful north-west American conifers, proved an extremely rapid starter on a variety of sites. On all the less fertile it flattered to deceive, and was

moribund within twelve years or so. It however persisted on good mineral soils, and proved of considerable interest because of its ability to root deeply on clays, with possible long-term benefits to drainage. It is one of the few broadleaved species which appears to have any chance of playing a useful part in upland spruce forest. On the whole, the exotic birches have not seemed likely to outrange the native species, but our arboreta contain several excellent American and Asiatic species, and one would not dismiss the possibility of finding useful hybrids for forest conditions.

Britain followed certain European countries in the vogue for the American Red oak (*Quercus borealis*), which has the advantage of growing fairly well on lighter and less fertile soils than required by the native oaks; it is also faster starting. It has now reverted to ornamental planting. No other exotic oaks have been seriously considered, though the good performance of the long introduced Turkey oak (*Q. cerris*) encouraged a check on its timber properties, in which Princes Risborough found it to be inferior (if possible) to its reputation. It is, however, quietly extending its range in our woodlands by natural regeneration.

Very late introductions to Britain were the Chilean *Nothofagus* species, of which *N. obliqua* and *N. procera* appeared in forest plots before they had become common objects in arboreta. Their initial success was such that Research took steps to have a wide scatter of plots established throughout the country. The interesting fact about these two *Nothofagus* species was that here we had broadleaved trees growing at coniferous rates on sites outside the normal productive range of broadleaved species in Britain. Their promise has been confirmed, though both have shown themselves prone to troubles from winter cold on certain testing sites (Low, 1967; Nimmo, 1971). Seeding of *N. obliqua* is prolific, and it seems possible that this tree may naturalise itself.

Another genus which has attracted serious attention only recently is *Eucalyptus*. A few private collections in favoured localities had shown that a small number of species could be grown in Britain, and unsuccessful plantings elsewhere had always indicated that winter minimum temperatures were the limiting factor for eucalypts in Britain. However, with better attention to seed source through the assistance of Australian experts (notably Martin of Tasmania), it has become apparent that many more species can be grown along the western seaboard, and it is possible that the range may be rather less restricted than has been thought. Some 36 species are now growing in the Forest Plots at Kilmun, Argyll. The acreage open to eucalypts in Britain must however be quite restricted, though it is much larger in Ireland. The only plausible motive for growing them would be for pulp, and if the cultivation of hardwoods purely for pulp is not now an economic proposition, it may yet prove useful to have gained experience of a number of fast growing broadleaved trees such as the eucalypts, alders, *Nothofagus* spp, and poplars. (Plate 41.)

Poplars and Elms

These two genera require special mention because of the weight of work put into them, and also because poplar and elm are normally grown "outside of the forest" on land of agricultural quality. A further point in common is that our work on them has stemmed from T. R. Peace's interest in their important specific diseases: bacterial canker of poplar and Dutch elm disease.

Before the war some useful work had been done on the nursery cultivation of poplars, but reliable establishment methods had not been worked out; though Peace was interested in clonal variation to bacterial canker some years before the war, and by 1939 some important introductions had already been made (e.g. "Androscoggin" in 1937). Peace commenced work on poplars as soon as he had moved to Alice Holt, concentrating on the collection of useful clones and the laying down of comparative trials in various parts of the country. He had valuable contacts on the continent with pathologists and poplar experts, and the International Poplar Commission provided a clearing house for information. J. Jobling was appointed in 1950 to work on the silvicultural aspects of poplars, and continued this work (amongst other activities) to the end of the period.

Peace's first approach to testing clones for susceptibility to canker was to expose plants to heavy natural infection from cankered trees established as sets from highly infected plantations. He found this unsatisfactory and moved to inoculation with bacterial slime from naturally occurring cankers. Most of the selection of resistant clones was done by this method, though later the refined method of Ridé using inoculum from pure culture was introduced by Burdekin.

A considerable number of "hybrid black" poplars derived from crosses between the European *Populus nigra* and the American *P. deltoides* were present in Britain. Sorting these out was a matter of some difficulty and P. G. Beak of Oxford did valuable work on the taxonomic side. The establishment of stool beds of authoritatively identified material was an essential step, and from 1949 onwards cuttings of accredited material were supplied to the nursery trade and interested individuals from the Commission's stool beds. The first selections were all old hybrids of repute in Europe which had been seen to perform well in Britain. From 1949 also the importation of poplars was governed by an order restricting such imports to clones of known canker resistance.

Somewhat later, interest began to centre on the balsams and hybrid balsams, especially the northwest American *P. trichocarpa* and its hybrids. (Augustine Henry had used *P. trichocarpa* as the pollen parent of the vigorous hybrid *P.* x *generosa* in 1912). The original introduction of this species by F. R. S. Balfour (which appeared to be represented by a single clone) had grown extremely well, but was canker-susceptible. Several introductions from American and Canadian selections were however resistant or much less susceptible, as were crosses with another American species *P. tacamahaca*. Some of these clones have now been approved for general cultivation. An interesting hybrid between *P. trichocarpa* and the Japan balsam *P. maximowiczii*, originating from an American breeding programme and registered under the clonal name "Androscoggin", attracted much attention from its extremely fast growth and good stem form in trials; unfortunately it proved insufficiently resistant to canker. Since so much breeding work was being done in Belgium and elsewhere on the Continent, it was not the policy to breed poplars in Britain, but this hybrid appeared so outstanding that it was considered worthwhile to re-create it using *P. trichocarpa* parents of known resistance. This was successfully accomplished by Jobling and Burdekin; the seedlings being tested for resistance by inoculation at an early age. Several seedlings showed promising degrees of resistance, but have since proved suspect.

In general, the important group of aspens and aspen hybrids (*P. tremula*, *P. tremuloides*, etc.) have proved difficult in cultivation, which is a pity as their suckering habit makes for cheap reproduction and, like the balsams, their soil requirements are more modest than those of the "blacks".

Poplar trials well scattered over the country have shown very clearly that *P. trichocarpa* and its hybrids have a wider climatic range than the blacks; these balsams are notably better suited than the blacks to the north and west. (Plate 40.)

Silvicultural work on poplars entailed a comprehensive and straightforward experimental enquiry into all aspects of their cultivation; nursery methods; planting and tending; spacement and increment; and pruning intensity. This fast growing clonal material is extremely responsive to treatment, and clearcut evidence has been obtained on all important topics. *Poplars* (Bulletin No. 19) by Peace (1952) provided an excellent background to the cultivation of poplars in Britain; it is brought up to date on certain points in *Poplar cultivation* (Leaflet No. 27) by Jobling (1963), who also described the establishment experimental work in *Establishment methods for poplars*

(Jobling, 1960). Poplar is a specialised crop with somewhat limited outlets; one of the main markets —for matches—seems likely to be fully supplied by Bryant and May's own very successful plantations. It is possible that we have put the cultivation of poplars onto a sound basis only to find that their usefulness in Britain is very limited, but Miller (1972) and Stern (1972) have discussed the possibilities of close spacing for pulpwood. The best hope lies in the growth rates and adaptability of the balsams and balsam hybrids, since it is with these that a respectable acreage can be worked up should it appear profitable to cultivate for any new market, such as hardwood pulp. (Plate 4.)

Peace had been concerned with Dutch elm disease since 1928, a year after its discovery in Britain. In the course of his pathological studies he had looked for instances of natural resistance, but the most promising apparent cases identified before the war did not stand up to test by inoculation, and the question of resistance in British elm populations remained open when he restarted work in 1947. By this time some of the early Dutch selections had shown much promise, and a programme of selection and breeding had been embarked on in the Netherlands. A few named clones were introduced to Britain for test in the late forties, and in 1954 a quantity of scions for grafting were sent to England by Heybroek who was leading the Dutch work at Baarn. It became the policy to collaborate fully with Heybroek, and scion material from British elms was also sent to him to assist in his breeding work. Elm being probably more important in Holland as an amenity and shelter tree (especially in the reclaimed polders) than as a timber producer, Dutch selections naturally emphasised the former qualities.

In 1959 elm was adopted as a tree for special study in the Silvicultural programme, largely because of the anxiety of the home timber trade over supplies. There was also concern over the possible disappearance of elm from the English hedgerows due to the disease, which is apt to become conspicuous after dry summers and good breeding conditions for the *Scolytus* beetle vector.

The programme commenced by surveying the material. The census of hedgerow and park timber (Forestry Commission, 1953) had shown the importance of elm as a "non-forest" timber source. There was however some doubt about the botanical status of the most valuable timber producing elms. Studies of elm populations by Richens (1955) had shown intergrading between indigenous species *Ulmus carpinifolia* (of the south-east) and *U. glabra* (the more generally distributed Wych elm). Elms such as the 'Dutch' and 'Huntingdon' elms appeared to have originated from hybridisation between *carpinifolia* and *glabra*. Experts in elm timber valued Dutch elm considerably higher than English elm (*U. procera*), but it was not easy to be sure that the timber merchant's Dutch elm equalled the botanist's *U. hollandica* var. *hollandica*. Hence in identifying desirable elms for propagation efforts were made to link selection with timber quality, sometimes by preserving scion material from elms which were due for felling and conversion.

The key to working up a clonal collection was a ready means of vegetative propagation, and the rooting of softwood cuttings in the mist-frame provided this, reasonably good takes being obtained after some development work on the method. Hardwood cuttings proved disappointing, and resource to grafting was only required as a preliminary measure to build up a stock for cuttings.

The primary test of elm selections is disease resistance, which is carried out by inoculation of the causal fungus. So far the Dutch ratings of resistance in their most promising elms have (as would be expected) been confirmed, but (as Peace thought) it does not appear that resistance is a common feature amongst native elms. There seems however to be promise of finding resistance in Huntingdon elm (*U. hollandica* var. *vegeta*).

Along with the selective side of the work, Jobling has also studied nursery and establishment methods for elms. The softwood cutting has been the basis, and cuttings rooted early in the summer in the mist-frame and lined out in the nursery for a further season have produced sizeable stock of good survival value. Very much larger stocks are obtained by relining for a further season, but the height advantage does not seem to be maintained in woodland planting. Elm plants are curiously unstable for a few seasons after planting, and so far no easy remedy has suggested itself.

The elms are up against many difficulties in Britain. The best opinion has been that Dutch elm disease will not by itself denude the country of elms. The English elm (*U. procera*) and *Ulmus carpinifolia* have considerable powers of regeneration by suckering. However, the English hedgerows are disappearing at an alarming rate, and these are the traditional sites for elms. No measures have yet been taken which are likely to encourage private individuals to plant elms, and it is not clear how the products of Dutch breeding or English selective work are likely to be planted in any quantity, except possibly by local authorities in connection with large-scale planning. However, one should not undervalue the salvage aspect of the programme—the preservation of valuable elms which would otherwise be irretrievably lost.

Arboriculture

Since forestry became an organised professional activity in this country, the trend has been for forestry and arboriculture to pull apart. This is most noticeable in the field of exotics, where the forester has identified virtually all the trees that are likely to be of use to him, and is not now greatly interested in the arboretum as a general trial ground for new species.

However, as forest authority, the Commission retains more than a marginal interest in arboriculture, and as the amenity and recreational interest in woodland increases, it is to be expected that foresters will often have to look at tree growth from an arboricultural rather than a utilitarian aspect. There is no central body for arboricultural research; nor could there be; the interests being far too varied. Forestry Commission activities have necessarily been rather "bitty". However, since many tree diseases and pests know no boundaries, Pathology (especially) and Entomology have always maintained an interest in the troubles of decorative as well as economic species. T. R. Peace's well-known work *Pathology of trees and shrubs* (Peace, 1962) illustrates this.

On the silvicultural side the most important activity has been the management of two great arboreta; the National Pinetum at Bedgebury, Kent, and Westonbirt Arboretum, near Tetbury in Gloucestershire, which was acquired from the fifth Earl of Morley in 1956. Westonbirt, founded in 1829, has extremely rich collections of flowering and autumn colouring trees and shrubs; for example its collection of *Acer* is thought to be the best in Europe. Management of such a place presents numerous problems, and the Commission wisely appointed a very strong advisory Committee at the outset. One of the objects of management recommended, which is now being implemented, is to make the collections as comprehensive as possible in those genera which were already well represented and important arboriculturally, such as *Acer*, *Quercus*, *Betula*, and *Magnolia*. Bedgebury, which was for many years a joint Kew/Forestry Commission enterprise, is now run by the Forestry Commission with an advisory committee. These two advisory committees have given us close relations with some of the most distinguished botanists and arboriculturists in the country.

An interesting recent development is the extension of the provisions of the Plant Varieties and Seeds Act 1964 to cover all genera with woody plants of decorative value. (The most important genus *Rosa* has of course been covered for some years). The Forestry Commission has agreed to play a part in providing space and other facilities for the necessary trial grounds and reference collections.

One very personal contribution to arboriculture should not go unmentioned. A. F. Mitchell has measured many thousands of specimen trees of an extremely wide range of species, thus providing a body of information on the growth rates of individual trees and, in many cases, close estimates of their ultimate dimensions. This information is of great use to planters. Several lists have been published (Mitchell, 1971; 1972 a and b) and many of the records will be found in the revised edition of Bean's classical work *Trees and shrubs hardy in the British Isles* of which the first volumes are beginning to appear in print.

REFERENCES

ALDHOUS, J. R., and LOW, A. J. (1970). Minor species survey. *Rep. Forest Res., Lond.* 59–60.

CHRISTIE, J. M., and LEWIS, R. E. A. (1961). *Provisional yield tables for Abies grandis and Abies nobilis.* Forest Rec., Lond. 47.

EVANS, W. R., and CHRISTIE, J. M. (1957). *Provisional yield table for Western hemlock in Great Britain.* Forest Rec., Lond. 33.

FORESTRY COMMISSION (1953). *Hedgerow and park timber and woods under five acres*, 1951. (Census Report No. 2).

JOBLING, J. (1960). *Establishment methods for poplars.* Forest Rec., Lond. 43.

JOBLING, J. (1963). *Poplar cultivation.* Leafl. For. Commn 27.

LOW, A. J. (1967). Nothofagus in Scottish state forests. *Scott. For.* 21 (4), 218–221.

MACDONALD, JAMES, WOOD, R. F., EDWARDS, M. V., and ALDHOUS, J. R. (1957). *Exotic forest trees in Great Britain*, Bull. For. Commn, Lond. 30.

MILLER, W. A. (1972). Poplar prospects. *Scott. For.* 26 (4), 285–287.

MITCHELL, A. F. (1971). Recent measurements of big trees in Scotland. *Scott. For.*: Pt 1, 25 (2), 109–112; Pt 2, 25 (3), 206–212; Pt 3, 25 (4), 277–285.

MITCHELL, A. F. (1972a). *Conifers in the British Isles.* Bookl. For. Commn 33.

MITCHELL, A. F. (1972b). Conifer statistics. In *Conifers in the British Isles: Proc. R. hort. Soc. 3rd Conifer Conf., Lond.* 1970, 123–285.

NIMMO, M. (1971). *Nothofagus plantations in Great Britain.* Forest Rec., Lond. 79.

PEACE, T. R. (1952). *Poplars.* Bull. For. Commn, Lond. 19.

PEACE, T. R. (1962). *Pathology of trees and shrubs.* Oxford: Clarendon Press.

RICHENS, R. H. (1955). Studies on *Ulmus*. I. The range of variation of East Anglian elms. *Watsonia* 3 (3), 138–153.

STERN, R. C. (1972). Poplar growing at close spacing. *Q. Jl For.* 66 (3), 230–235.

STIRLING MAXWELL, SIR JOHN (1932). The influence of exotic conifers on silviculture in the British Isles. In *Conifers in cultivation*, Royal Horticultural Society, London, pp. 43–54.

WOOD, R. F. (1955). *Studies of North-west American forests in relation to silviculture in Great Britain.* Bull. For. Commn, Lond. 25.

Chapter 16

AFFORESTATION OF SPECIAL SITES

While much of the Commission's earliest work on establishment centred on major land types, there was always a body of work concerned with extending the boundaries of plantation. This usually took the form of trials planted beyond the then accepted limit of elevation, fertility, or both. Such trials often incorporated comparisons of species, manurial and cultural experiments, or they might simply be pilot plots established under the best known method. Many of them, being laid down in a period when advances in technique were coming into general practice, were surrounded by plantations before they had yielded their full evidence.

In the post-war period there was an urgent need to obtain more evidence on planting limits. This was particularly important in Scotland where very large acreages of land in the north and west are highly exposed, and also covered by types of peat which had not as yet proved a successful planting medium. In England and Wales the position was rather different; the north of England certainly possessed large tracts of doubtfully plantable land, but elsewhere such problems were dissected into local environments of no very great extent. The special question of high elevation of course affected Wales, but the areas in doubt were much less than those in Scotland.

Although no hard and fast distinctions can be made, it will be convenient to deal with special site work in this Chapter, leaving the principal experimental subjects (manuring, cultivation etc.) to be dealt with separately, since many of the developments in establishment technique spread themselves across the boundaries of types of land.

Northern Scotland

By far the greatest volume of work was concentrated in the far north, where the poor peats of Caithness, Sutherland and Wester Ross had been regarded as unplantable in the pre-war period. Much potential forest land was also highly exposed. A new series of trial plantations was started in the late forties and by the mid-fifties a large programme had been carried out. The trials took the form of plantations laid down by the latest technique—on peats this was plough draining using the new Cuthbertson ploughs. Lodgepole pine was the principal species, and others, including Sitka spruce, considered worth trials were incorporated in a matrix of this tree. Phosphatic manuring was of course general. Some of the plantations were made at the request of the Department of Agriculture as experimental shelterbelts or shelterblocks. In 1953 experimental plantings were made in Shetland, and a year later in Orkney.

This work established clearly enough that it was possible to establish plantations on any site which could be ploughed, but on the poorer peats no species other than Sitka spruce and Lodgepole pine showed any promise, and it became increasingly evident that on the least fertile peats the choice might well be restricted to Lodgepole pine alone. The trials also indicated that exposure was going to be *the* limiting factor, especially in the isles, but also on some mainland sites, and this became a special object of study.

Elevated and Exposed Sites

General meteorological evidence on prevailing wind velocities puts the most exposed British coastal districts almost in a class of their own; this was known, but there were not enough wind-velocity statistics to suggest where the limits for satisfactory tree growth might be, and the older plantations in exposed districts were not usually composed of the most wind-resistant species. Hence exposure had to be treated to start with on a trial and error basis.

A cheap and reasonably consistent physical measure of exposure was urgently required, and to a great extent this was provided by the development of the flag method (Lines and Howell, 1963) by Edwards, Zehetmayr and other research workers in Scotland, from the original idea of an Orcadian, who built his house at a point where a series of unhemmed flags tattered least. The development required the standardisation of the material for the flag, its dimensions, mounting and the method of assessing tatter. The factors governing the rate of wear of the flag are undoubtedly complex, and a simple correlation with run of wind not to be expected; but the rate of tatter of flags fitted in well with observations of the growth of trees (especially Lodgepole pine) under different degrees of exposure at the various northern trial areas. It therefore became possible to use flags for advance estimates of exposure on questionable sites, and even latterly to construct exposure "contour" maps. The flag method has run into difficulties at high elevations in Wales, where (perhaps surprisingly) icing seems to be more prevalent in the middle of winter than at similar elevations in Scotland. (Plate 42.)

A special exposure problem was encountered in the Pennine hills where industrial pollution from the West Riding of Yorkshire and Lancashire is a complication. In fact, an extensive series of trial plantations served to dismiss this region as an afforestation prospect. Lines (1961–64) studied the difficult question of separating exposure and pollution factors for a number of years, and gained much useful experience in gauging sulphur dioxide pollution under forest conditions.

High elevation in Britain is largely a matter of increased exposure so far as tree growth is concerned, and observations of high level trial plantations usually suggests that exposure limits growth before the reduction of growing season temperature is significant. Trials have shown that the planting limit varies very greatly, fair growth being obtained at elevations as high as 2000 ft in Central Wales and parts of the Scottish Highlands on the least exposed slopes, whereas in the northern isles and the Hebrides exposure may be limiting at sea level.

No trials have suggested that any specialised mountain trees have any value at high elevations in Britain, which is not surprising since our mountain climates are merely "bad oceanic" rather than continental/alpine. The flag method has been most useful in tracing the upper limits of planting, especially in extrapolating from existing trial plantations. It was, however, a growing concern during this period that we knew too little about British mountain climates. The Nature Conservancy's work in this field is now of great importance, and the Meteorological Office has recently paid much attention to the development of automatic recording stations.

The general subject of wind and its effects on tree and other plant growth has embraced more work than can be mentioned here. Stability against windthrow is dealt with in Chapter 21. But all work on wind, whether the interest is shelter, the stability of crops, or growth, has a common concern in aerodynamics. The first scientific studies of shelterbelts in this country were undertaken by J. M. Caborn of Edinburgh University from 1953 to 1955 with the assistance of a grant from the Forestry Commission. Caborn's *Shelterbelts and microclimate* was published as Forestry Commission Bulletin No. 29 (Caborn, 1957), and he has continued to work in this general field. Today Edinburgh has an experimental wind tunnel of very advanced design which will prove a great asset in the physiological study of wind on plant growth. The subject is of course common ground between agriculture and forestry.

Other Special Sites
A number of sites which attracted individual studies might be described as pockets of resistance; they were not parts of the general frontier of plantability, and most of them answered readily enough to advances in cultivation and manuring. The steep grasslands on shaley soils in Central Wales did not prove difficult once it was found that some of them were extremely phosphorus deficient, though on the steepest slopes ploughing was not easy, and there appeared to be a case for herbicides for grass control in the establishment stage. Some of these soils were far from ideal for Sitka spruce, and as a number of other trees were tried at the main experimental area, Taliesin, in 1956; *Tsuga heterophylla*, and *Abies nobilis* showing considerable promise. The hybrid cypress x *Cupressocyparis leylandii* also made remarkable early growth on this elevated site. Work was also done on several southern heaths, the most interesting site being Goonhilly Down, on the Lizard Peninsula, Cornwall. This is one of the few Serpentine formations in the country, and American accounts led us to expect nickel toxicity. However, the soils turned out to be derived in large degree from loess of granitic origin; its reaction is near alkaline, due not to calcium, but magnesium, and another interesting feature is that the dominant heath is the very localised *Erica vagans*. Since a substantial area was available for planting, trials were started in 1955 prior to further acquisitions. In the event, no toxicities appeared, and the most notable feature was the extremely large response to phosphorus. In this climate the species choice was wide open, but coastal exposure handicapped many subjects. *Pinus radiata* proved easily the most vigorous species.

The Chalk is worth passing mention. The formative work had been done before the war, and with the revival of downland agriculture little more of this type came in for planting. The most important advance was the appreciation of the value of superficial acid layers in averting lime induced chlorosis in susceptible species. This led to the idea of subsoil cultivation with minimum surface disturbance (i.e. without share). Grass competition being more serious on chalk downlands than on any other site, the herbicides showed to great advantage here, especially the graminicide dalapon.

The Commission's few duneland sites had never attracted much research in *establishment* methods (Culbin, Moray, was an important area for soil nutritional and physical studies), sand fixation and planting methods being derived directly from classical European work. A small amount of work was done in the fifties at Newborough Warren, Anglesey, on bituminous emulsions for sand fixation and covering of direct sown patches. This was successful up to a point, but young seedlings were not able to withstand the environment in their first winter, and hence the method showed no promise of replacing traditional methods employing shelter. The technique has of course been much employed in fixing and vegetating steep slopes in road works.

An assortment of industrial sites received attention in the period. One of these, the opencast ironstone workings of Northamptonshire, had at one time prospects for large-scale planting, since before the enactment of the Mineral Workings Act in 1951 governing the restoration of such workings, it was the practice to leave the land after extraction of the ore in "hill and dale" formation. One of the largest firms had planted a considerable area of such hill and dale in the Corby area, and other such sites had been planted by private estates. A survey of such plantations was made by R. D. Pinchin (1952), but no experimental work was undertaken, since the policy of restoration for agriculture intervened. Hill and dale looked like a gigantic ploughing operation, mixing (more or less) all the strata of the overburden. Planted as soon as possible after the working ceased, some excellent plantations resulted, especially on the lighter overburdens. Plate 43 illustrates afforestation at opencast coal workings in South Wales.

Work done at Wareham Forest, Dorset, in 1960 on hill and dale formations of disused gravel workings disclosed no special problems. Several species survived well, and Corsican pine, *Pinus radiata* and Leyland cypress displayed appreciable vigour. Application of P significantly improved growth rates, but K had less effect.

Other industrial sites were not forestry prospects, but of importance in the local amenity. Plantations on deep mined spoils, principally colliery wastes, were surveyed by C. V. Thirgood (Wood and Thirgood, 1955) in 1954. Many successful plantations were found (except on modern tall conical "bings"), and it was observed that, as with opencast spoils, there appeared to be little or no advantage in waiting for weathering or ecological succession to advance. Later, Neustein in Scotland carried out establishment experiments on the burned shale bings in West Lothian, these being residues of the old shale industry. Most of these various spoils have proved to be better sites for tree growth than one would expect. Of the many trees tried on them one might single out the alders as the most successful. Industrial waste planting is of course often subject to the hazard of pollution, and sometimes soil toxicity.

One curious substance appeared late in the period, when the question arose of planting trees on deposits of pulverised fuel ash (the waste of modern large coal-fired power stations). Large quantities of this material were being dumped in old quarries or artificial lagoons, and covered with a thin layer of soil to stabilise the surface. Apart from its odd physical nature, being comprised largely of minute glassy bubbles, "PFA" contains boron at a level toxic to most plants. Reading University conducted basic research on PFA in relation to plant growth, and some small-scale pilot plantings were made on deposits of it at one or two sites. Common broom and alder grew quite well on it. The substance has fortunately however found a number of industrial uses, and is receding in importance as a waste material.

REFERENCES

CABORN, J. M. (1957). *Shelterbelts and microclimate.* Bull. For. Commn, Lond. 29.

LINES, R. (1961–1964). Experiments on atmospheric pollution and artificial shelters. *Rep. Forest Res., Lond.* 1960, 40; 1962, 40–41; 1963, 37–38.

LINES, R., and HOWELL, R. S. (1963). *The use of flags to estimate the relative exposure of trial plantations.* Forest Rec., Lond. 51.

LINES, R., and NEUSTEIN, S. A. (1966). Afforestation techniques for difficult sites—wetlands. *Scott. For.* **20** (4), 261–277.

PINCHIN, R. D. (1952). Survey of plantations on opencast ironstone mining areas in the Midlands. *Rep. Forest Res., Lond.* 1951, 51–54.

WOOD, R. F., and THIRGOOD, J. V. (1955). *Tree planting on colliery spoil heaps.* Res. Brch Pap. For. Commn 17.

Chapter 17

NUTRITION OF FOREST CROPS

The value of phosphatic manuring had been well established in pre-war experiments on the heaths and the infertile peats. The range of conditions under which useful responses to phosphorus might be expected was not however known with sufficient accuracy, nor had the application of phosphatic manures at planting become general practice on any major land type. Basic work on forest soils related to nutrition had been very limited before the war. Aberdeen University had been concerned from the outset, and later the valuable association with the Macaulay Institute had been established. In the post-war period the position was different, since several other important institutions took up work on forest soils; notably the Commonwealth Forestry Institute, Oxford; Rothamsted; and the Nature Conservancy.

The Commission's own experimental work in this field expanded greatly in the post-war period. Much of it followed the pre-war lines of empirical experiments in manuring crops at the establishment phase; usually associated with cultural operations such as ploughing and draining; but by the end of the period the nutrition of crops of a wide age span was under investigation. With the establishment of the Soils Sub-Section in 1956 analytical approaches for soils and plant material became available, and the subject ceased to be purely empirical.

In the immediate post-war years, work on the important heathland and moorland types paused somewhat for the digestion of the great weight of pre-war and wartime experimentation, leading to the publication of Zehetmayr's two Bulletins. In the early fifties however work on establishment problems was again very active, and by this time Crowther had several seasons of experience of manuring in upland and other heaths in connection with the testing of plants raised in his nursery programme. He had compared a number of forms of phosphatic fertiliser, and had found no form superior to superphosphate, which did not reproduce the pre-war troubles associated with superphosphate in the mixture with ground mineral phosphate known as Semsol. Crowther also obtained early responses to potash and nitrogen on certain heaths, the nitrogen effects being occasionally quite large. These elements however were to play little part in *establishment* work, since the responses tended to be ephemeral. Indeed on many of Crowther's sites, which were usually *Calluna* dominated, the mode of cultivation of the period (single-furrow ploughing) was inadequate to bring Sitka spruce into canopy before typical *Calluna* check set in. The excellent growth of fully manured seedlings before the reinvasion of *Calluna* however provided a new standard.

In the *Report on forest research* experimental work is usually classified by the stage of the crop, i.e. (i) establishment, (ii) checked or slow-growing crops before canopy closure, and (iii) pole stage or older crops. In establishment work manuring was frequently at standard rates, the main interest being the methods of cultivation or drainage, or the choice of species. Work was done on forms and rates of phosphatic manure, following Crowther. Scotland had settled on ground mineral phosphate after the early work on basic slag, but other forms were studied, especially triple superphosphate, which was attractive from the weight aspect (Edwards, 1959). Triple super performed well and gave no trouble. Work was also done on placement. The relatively insoluble GMP could simply be "thrown" at the planting point but, following Crowther again, the practice was to place more soluble forms, such as super and triple super, in bands a few inches from the point of insertion of the plant. Crowther obtained no immediate advantage from inserting pellets of phosphate beside the plant, and generally speaking there has never appeared to be a better way of applying phosphate than superficially adjacent to the plant. The possibility of placing GMP under the ridge during the act of ploughing was however examined in the early sixties.

It is remarkable that on so many sites initial responses to phosphorus confirmed the conventional rates established before the war, roughly equivalent to 25 lb P per acre. In fact foresters got used to thinking of phosphorus as a catalyst or root-growth stimulator rather than a nutrient. However, in the late fifties and early sixties it became apparent that certain very phosphorus-deficient soils such as those on the Culm Measures, the Serpentine, and some of the Tertiary formations in southern England, required dosages of twice or three times the normal for sustained response.

Forms of phosphorus never gave any very clearcut results. The weight of the evidence supports Atterson's comment that it is the amount of P per unit cost which ultimately matters, and this has favoured the less processed forms such as GMP.

Important potash deficiencies were noted for the first time on deep peats in Scotland in the mid-fifties and it later became apparent that certain Welsh peats were particularly potash deficient. With these exceptions potash has not been a critical element in the establishment phase. There is usually a sufficient release of nitrogen in the act of cultivation, and subsequently invading vegetation is apt to take up the lion's share of applied nitrogen, hence it has not generally been found useful to apply it at planting. But there has been a growing interest in full fertilisation on very exposed sites, where both potash and nitrogen appear to render the plant less susceptible to blast damage, and possibly also to frost. "Luxury" nutrition has been tried in a number of pilot plots at high elevations or other exposed sites.

The great majority of manurial experiments concerned with establishment have been carried out on sites where there was reason to expect a response from fertilisation, but phosphate has frequently been applied (experimentally and otherwise) on sites where no or little response was obtained. A survey of trials of seedling stock planted widely throughout the country afforded an opportunity to relate response to soil phosphate content; it was found that there was a fairly clear indication of a threshold value of 1,300 parts per million *total* phosphate above which little or no response could be expected, but below which response would be the rule. (Green and Wood, 1957). The vegetation has always been a useful guide to the experienced man, but in the early post-war period a number of instances were found where an apparently innocent flora concealed serious phosphorus deficiencies; this was very notably the case in some of the grassy Welsh uplands on Ordovician and Silurian shales.

It had been appreciated before the war that there were conditions under which it was easy enough to establish (or rather start) Sitka spruce by suitable cultural methods and phosphatic manuring, but which did not permit satisfactory growth or even the closure of canopy without further intervention. Most of these troubles were associated with the return of *Calluna* on both heathland and moorland sites. Underlying the direct competitive role of *Calluna* were the basic soil properties, physical and chemical, and much of the nutritional work was directed towards sorting out this complex. Early in the period Oxford had started work on the various factors influencing tree growth on the heaths, most of the work being centred on Allerston Forest, Yorkshire. Leyton (1950) suggested at an early stage that the primary role of *Calluna* in checking spruce was by inducing nitrogen deficiency. Handley (1961) undertook a profound study of the biochemistry of mortype humus by *Calluna*, and heathy vegetation in relation to nutrition and early growth became a dominant theme in experimental work over much of the post-war period.

In the early fifties the assumption was that Sitka spruce which had gone into deep check following *Calluna* reinvasion could not be brought out by manurial applications alone. Effective experimental techniques such as suppression by mulching or interplanting with pines or broom were not practicable on any scale. However, some interesting exceptions began to appear in Holmes' work on checked crops in the south-west of England and in Wales. Wilsey Down in Cornwall had spruce crops dating from the thirties, most of which were in deep check in a heathy vegetation over very infertile Culm Measures Early experiments introduced new species in broom nurses with phosphatic manuring. Phosphate responses were good, though it was noticed after a season or two that the effects of standard rates fell off rapidly. Somewhat tentatively, manuring was extended to the original checked crop, which was to all appearance moribund. The results were quite remarkable, since wretched plants with few live needles and no leaders "came alive" and started to grow. It was found that broadcast dressings were more effective than placed applications; not surprisingly since the checked spruce were subsisting on extremely long lateral roots running just below the humus layer of the ground vegetation. The response was entirely to phosphorus. Like many other western heaths, Wilsey Down carried the dwarf gorse *Ulex gallii* as a component of the vegetation, and it was an obvious hypothesis that the presence of a leguminous shrub might distinguish such sites from typical *Calluna* heath in respect of nitrogen nutrition. Analyses of Wilsey soils showed phosphorus contents a good deal lower than those of most mineral soils of which the Commission had experience, but later it was appreciated that soils over a number of other southern formations such as the Serpentine, the Bagshot beds, and some of the Wealden series were equally deficient.

The success of hand applied broadcast dressings at Wilsey led to larger-scale machine applications by tractor-mounted spreader (the "crop" representing little obstacle in the poorer compartments), but for most of the area it was plain that aerial application was the only practical method. This was carried out by fixed-wing aircraft in 1959 over some 100 acres of the poorest crops. Three cwt of triple-superphosphate per acre was applied at an approximate cost of £6 per acre. This was the first British experience of aerial fertilisation of forest crops, a technique which had been applied on a large scale for the improvement of pasture in Australia and New Zealand, and had been pioneered in forestry by the Swedes. Wilsey was an exceptional place in many ways. Responses to phosphorus increased to quite unusual rates of fertiliser. One of the deficiency symptoms was peculiar; growing shoots wilting quite suddenly and irreversibly. Seen a week or two afterwards, the symptom could easily be mistaken for frost damage, but whatever the physiological mechanism, it did not occur with adequate phosphatic manuring (Holmes and Cousins, 1960). (Plates 44, 46, 47.)

Wilsey experiences encouraged work on checked or slow-growing crops on a number of other types of site in England and Wales. Close analogies to Wilsey were found in some of the mid-Wales forests, notably Tarenig, where much spruce grew very slowly or checked on steep grassy slopes over shaley soils. Dwarf gorse was usually present and, as at Wilsey Down, good responses were obtained to broadcast phosphate. On some of the poorest southern heaths on the Bagshot beds even pines were found in check. Here the nitrogen story was important, and in the presence of phosphate considerable nitrogen responses were obtained, and the responses to both elements were stepped up by the removal of *Calluna*.

Many examples of check could be ascribed to inadequate site preparation, failure to manure at planting, or wrong choice of species. A more fundamental problem was to maintain growth of Sitka spruce on the poorest peats, and it became increasingly clear that this was primarily a nitrogen question. It was observed that Welsh peats appeared to be better in this respect than Scottish peats of apparently similar character (Binns, 1967).

Peat analyses showed very large differences in phosphorus and potash contents between basin bogs, blanket peats, and raised bogs, in descending order. On the poorest deep acid peats in Scotland the sole choice of species appeared to be Lodgepole pine with PK manuring; no practical programme of fertilisation being as yet available to meet the nitrogen requirements of Sitka spruce, or change the nitrogen cycle in its favour.

Work on the manuring of pole stage and other crops in canopy followed naturally enough from success in dealing with checked crops. Worthwhile responses to nitrogen in semi-mature stands had been obtained on the Continent, especially in Sweden. Rennie (1955), in his studies of nutrient uptake and soil reserves of the major elements, had suggested that nutrition would severely limit tree growth on the poorer heaths, even in crops which had been successfully established and had closed canopy.

The first experiments were laid down in the mid-fifties on low quality class pine stands in England, and work was later extended to spruce stands of various quality classes. The earliest experiments were ambitious in design with three levels of five fertilisers (N, P, K, Ca, and Mg) in a factorial design with partial replication. Vernier girth bands were used to obtain early information on responses. The experiments were tedious to lay out and assess, and later experiments were much simpler when it became clearer what elements were of interest on particular sites. This work continues, and has not as yet established a very clear pattern of response. So far the most important responses have been to P and N. Some spruce stands have given worthwhile responses to P when on rates of growth this would scarcely have been expected. Other spruce stands have not responded, and in certain cases this may have been due to poor physical conditions. The most consistent results have been with pines in the north-east of Scotland, where J. B. Craig and H. G. Miller (1966) of the Macaulay Institute conducted a number of experiments on nitrogen nutrition of pines on sandy soils in collaboration with the Commission. At Culbin, Moray, the experiments were part of a general enquiry into nitrogen in the ecosystem, and the interesting fact emerged that organic material in the soil and litter contained over 200 times as much nitrogen as the average annual uptake of the crop, which itself was very deficient, as shown by the greatly increased uptake and increment from added nitrogen. Binns (1962) showed that in the peat ecosytem the total amount of N is two or three times that at Culbin. The mobilisation of nitrogen in forest soils remains one of the great problems of the cool/temperate regions.

The Macaulay Institute's work at Culbin, started by J. D. Ovington, may be regarded as part of the general enquiry into the effects of plantation on the site, and more will be said about this in Chapter 24. Narrower and very practical questions were concerned with deficiencies in foliar contents, the uptake of nutrients, and the distribution of added nutrients in the tree, soil, and competing vegetation. W. H. Hinson followed Macaulay procedures in setting up the analytical facilities at Alice Holt in 1956. Foliar analysis developed quickly into a useful diagnostic tool, and by the end of the period the Alice Holt laboratory was handling 3,500 foliar samples each year, in addition to a smaller number of soil analyses. The analyses of peats were conducted by the Macaulay Institute for most of the period, but recently arrangements were made with the Soil Chemistry Department of the Edinburgh University School of Agriculture for extensive analyses of important peat types. By the end of the period virtually all important projects for refertilisation of slow-growing crops were supported by analyses.

Deficiency symptoms in older plants are not usually as graphic as those of Sitka spruce seedlings (well portrayed in Bulletin 37). To study conifers with known shortages of major elements, a type of experiment known flippantly as a "deficiency garden" was laid down at Wareham on an extremely infertile site in 1960. The idea was to induce deficiency symptoms by applying all but one of each of the major elements in turn, under weed free conditions, and with great care to avoid the spread of added fertilisers. The approach was later repeated on peat at Eddleston in Peeblshire. Colour symptoms were useful, but not so individually diagnostic as those of one-year-old seedlings, the range being limited to yellows. Certainly at Wareham, the main interest of the experiment lay in the demonstration of growth on the plots with full (repeated) fertilisation under weed free conditions. Though this was not a condition which could be appreciated in practice, it had value as an approximation to the optimum growth to be expected of the general environment, a measure (largely of climate) about which there is extreme doubt in many circumstances (Binns and Mackenzie, 1969).

The most obvious advance over the post-war period was the increased use of phosphatic manuring at planting, but it must be admitted that it entered into general practice rather too slowly on sites in need of it. In certain cases this was due to lack of Research coverage on the sites concerned, because they had not been recognised as presenting any difficulties. Elsewhere neglect was less excusable because the evidence was readily available However, by the early sixties it might be said that all important sites on which phosphatic manuring was critical for establishment were receiving it, and probably the great majority of sites on which a worthwhile response was to be expected also received phosphates. Where phosphorus is critical the only economic argument is whether to plant the site or no, but phosphorus has also given a range of growth responses on sites where a crop would have been obtained without it. Probably the largest and most lasting effects have been with pines on heathland sites. Neustein (1963) gives an example where the effect of phosphate applied at planting to *Pinus contorta* amounted to a 50 per cent increase in the basal area of the crop at 30 years of age.

Nitrogen and potash belong almost entirely to the post-war period, and the limits of their use have probably not been reached. One of the largest gains of the post-war period has been in the diagnosis of deficiencies in slow growing or checked crops, and this has given a great measure of confidence in applying remedial measures.

It is probable that we have approached the limits of what can be done by manuring in the establishment of the crop, but on the other hand much remains to be learned about fertilisation of the growing stand. So far, practically the entire effort has been put into pure crops, and perhaps more should be done with subsidiary species in mixture or rotation, especially with a view to improving the nitrogen regime.

REFERENCES

ATTERSON, J., and DAVIES, E. J. M. (1967). Fertilisers —their use and method of application in British forestry. *Scott. For.* **21** (4), 222–228.

BINNS, W. O. (1962). Some aspects of peat as a substrate for tree growth. *Irish Forestry* **19** (1), 32–55.

BINNS, W. O. (1967). Manuring of young crops on peat soils. *Rep. Forest Res., Lond.* 1966, 32–33.

BINNS, W. O., and MACKENZIE, J. M. (1969). Nutrition of forest crops ("Deficiency gardens"). *Rep. Forest Res., Lond.*, pp. 66 and 68.

CRAIG, J. B., and MILLER, H. G. (1966). Research on forest soils and tree nutrition. *Rep. Forest Res., Lond.*, p. 95.

EDWARDS, M. V. (1959). Use of triple superphosphate for forest manuring. *Rep. Forest Res., Lond.* 1958, 117–130.

GREEN, R. G., and WOOD, R. F. (1957). Manuring of conifer seedlings directly planted in the forest. *Rep. Forest Res., Lond.* 1956, 132–139.

HANDLEY, W. R. C. (1961). Further evidence for the importance of residual leaf protein complexes in litter decomposition and the supply of nitrogen for plant growth. *Pl. Soil* **15** (1), 37–73.

HOLMES, G. D., and COUSINS, D. A. (1960). Application of fertilisers to checked plantations. *Forestry* **33** (1), 54–73.

LEYTON, L. (1950). *The growth and mineral nutrient relations of tree growing in Calluna dominated sites.* Paper for 7th Int. bot. Congr., Stockholm.

NEUSTEIN, S. A. (1963). The effect of phosphate, applied at planting, on a crop at the pole stage on upland heath. *Rep. Forest Res., Lond.* 1962, 141–145.

RENNIE, P. J. (1955). The uptake of nutrients by mature forest growth. *Pl. Soil* **7** (1), 49–95.

Chapter 18

CULTIVATION AND DRAINAGE

By the end of the war ploughing for drainage and cultivation was an established practice, though by no means general. It is now general; at least 70 per cent of the annual area afforested being ploughed for one or other objective. This has come about by the evolution of ploughs and techniques over the whole period. Current ploughing practice has been described by Taylor (1970) in the Commission's Forest Record No. 73, in which full descriptions of ploughs and their work on different types of site are given. It is not possible to deal adequately with the development of this type of machinery (or indeed any other mechanical development) in a general account of research, and only the trend of progress can be described.

The Research part of the developments in ploughing techniques has, in the main, been concerned with specification rather than design. Experimental results have been reinforced by basic studies of soils and rooting habits; leading to concepts of what a machine should be designed to do. The actual design has been a matter for the Commission's Mechanical Engineers and the few firms who have specialised in this type of machinery. In particular the Scottish firms of J. A. Cuthbertson Ltd, Biggar, Lanarkshire, and W. Clark and Sons of Parkgate, Dumfries, should be mentioned for their outstanding work in this field; and if it is not invidious to single out the name of one of Commission's engineers, it should be that of the late J. W. Blaine. (Plate 54.)

The main developments in traction have been in the successive introductions of more powerful machines, with adaptations of tracks and linkages, rather than on the design of tractors for this limited field of use.

Zehetmayr's two Bulletins, Nos. 22 and 32 (Zehetmayr, 1954; 1960), dealt comprehensively with establishment problems on the peats and upland heaths respectively up to the middle fifties. These have of course been the two great fields of experiment, but ploughing for drainage and cultivation has been carried out to a lesser extent on other types. There is no hard and fast line between cultivation and drainage; most efficient cultural operations do both. The distinction is one of emphasis; but since research work tends to concentrate on conditions where one form of improvement is the primary requirement, it is convenient to discuss the two topics separately. In all research work on site preparation the post-war period showed increasing concern with the later growth and yield of crops, and hence there is a marked stepping up in the intensity of experimental treatments.

Cultivation

On heathlands, spaced furrow ploughing with the RLR plough, and from the early fifties the Tine plough, was the normal method of cultivation. There were however strong indications from pre-war experiments that complete ploughing was advantageous on certain heaths; there was also good reason to study greater depths of cultivation. The large and well designed experiment laid down at Teindland, Laigh of Moray, in 1952, and repeated at Harwood Dale, Yorkshire, in 1954, was the most influential piece of work in the period. Briefly, the experiment compared depths and intensities of cultivation in various combinations, providing a wide range of soil disturbances. Owing to the excellence of its design and concept it has given increasingly valuable and significant information for eighteen years. This experiment and its results are described by Zehetmayr (1960), Henman (1965) and others. It showed that there was a close relationship between growth and the volume of disturbed soil, especially on sites where, following the breaking up of compaction or pan formation, there was reasonably free water movement through the profile. The most intensive treatments in the experiment, such as deep complete ploughing, were regarded at the time as uneconomic; recent estimates of yield based on height growth however suggest (to say the least) that the question is open. (Plate 45.)

At the same time a study of the rooting patterns of various species on heathland soils in Scotland and Northern England was being conducted by C. W. Yeatman, who dissected numerous root systems of trees growing in experiments comparing different methods of cultivation. He found that on compacted heathland soils rooting in the absence of cultivation was virtually confined to the thin surface peats scarcely penetrating the leached A horizon. Following cultivation, roots freely penetrated disturbed soil, but were very much restricted to the disturbed zone. Hence all partial methods of cultivation produced asymmetrical root systems. Yeatman's studies were published as Forestry Commission Bulletin No. 21 (Yeatman, 1955).

Experiments similar in approach to the Teindland Expt 81 were later conducted on a variety of sites, including a number of southern heaths, and some of the compacted Welsh mountain soils. The "hard" heaths of the north-east of Scotland provided a special case, since these had indurated till sub-soils, which constituted a much greater restriction on rooting than podsol pans. Late in the period attempts were made to break such indurations using very heavy rippers designed primarily for opencast mining.

Some attention was given to machines other than ploughs for complete cultivation. Discing, even with very large machines, did not show much promise, and a scaled-up type of rotary cultivator appeared too slow and costly.

Spaced furrow ploughing on freely draining soils offers several positions of planting in relation to the ridge and furrow. From the introduction of ploughing the commonest practice was to plant on the ridge, and with somewhat superficial ploughing and incomplete rupture of the pan, the furrow (even on heaths) might be very wet in winter. With the introduction of the tine plough subsoil cultivation below the furrow effectively drained this position, and it became a possible planting position. Scottish experiments in the late forties and early fifties indicated that for survival the tine slit in the furrow was in fact the best planting position, compared with the ridge, and the edge of the furrow beside the ridge, though this latter position gave the best early growth. Planting in the tine slit became common practice, but in the early sixties it was observed that quite young pines tended to be unstable in high winds when planted by this method. Root studies showed poor, unilateral systems, with little development at right angles to the ploughing direction, whereas trees planted on the ridge or beside it had exploited a considerably greater volume.

Generally speaking, the results of most work on cultivation have been clear and the explanations (in terms of the discouragement of competing vegetation and the provision of rootable soil) obvious enough. It is now technically possible to deal with vegetation control by chemical means (of which more will be said later), and this may occasionally modify the degree of cultivation necessary. But cultivation is primarily the increase of rootable volume, and it remains a very open question what degree of cultivation forest crops will stand, economically speaking. Much depends on the degree to which poor heathland types can be upgraded by cultivation, manuring, and correct choice of species, into permanently productive environments, and we are far from knowing all the answers yet.

Drainage

British forest soils (or potential forest soils) which are *not* free draining through the profile are of far greater extent than those that are, and with the concentration of afforestation since the war in the wetter upland regions, the topic of drainage has assumed even greater importance. In the very wide range of work which has been undertaken since the war, one might single out the interaction of the crop and the drainage system as the most important theme.

Most pre-war experiments had been concerned with the development of the turf planting method, and hence comparisons of intensity of drainage were more enquiries into turfing systems (including the cultivation element) rather than strict drainage experiments. By foresight a few experiments compared depths of main drains to depths of 30 ft; well in advance of equipment to render such operations practicable.

In the Border forests, where large acreages had been established on "easy" *Molinia* peats over tills by shallow draining and turfing, it was noted (Macdonald, 1950) that spruce root systems were extremely superficial, and often crossed the drains. Drainage experiments of the late thirties on this type showed responses to depth and intensity, and preliminary studies in drain deepening in spruce stands in the Borders showed that water levels in test wells were very markedly influenced by deepening drains to 30 in, an encouraging indication that these clayey tills were in fact "drainable" (Stewart, 1959). The stability of crops on the Border type was much in mind when starting this work; in fact Macdonald proposed to test the effects of draining (and thinning) by measuring the pull required to uproot sample trees, a technique which was developed a good deal later by A. I. Fraser.

The introduction of the Cuthbertson ploughs in 1945, with speedy differentiation into single-throw ploughs designed for true draining and double-throw machines providing turf from shallow furrows, was one of the key developments in moist upland afforestation. Once adequately powered and wide-tracked crawler tractors became available, turf planting evolved into a mechanised system. Drainage, however, remained the weak point, since there was extremely little new evidence available till the early sixties to support the Research view that drain depth and intensity was inadequate in general practice. By this time some of the earlier post-war enquiries had begun to yield results. One important indication had been obtained from a drainage experiment at Bernwood Forest near Oxford, where new plantations had been made after clearance of coppice on heavy Oxford clay—a soil type which was often considered undrainable. Levels in test wells showed considerable differences in behaviour according to drainage intensity, and somewhat to most people's surprise, the new crop also responded within a few years of planting. This experiment (described by Fourt, 1961) was an incentive to thinking in fundamental terms about drainage of clay soils, with a better appreciation of the role of tree growth in drying out and fissuring the clays, such fissures remaining partly open in winter through the incorporation of organic matter and providing channels for lateral movement of water to the open drains. This drew attention to the possible advantages of trees, such as the alders, which are capable of deep rooting in clays.

The physical changes in peat on drying had long been of interest. G. K. Fraser of the Macaulay Institute demonstrated at Inchnacardoch that virtually no change occurred in twenty or more years when a small (tenth-acre) plot of peat was isolated by deep drains to the subsoil. Very different was the behaviour under nearby thriving crops of Lodgepole pine; marked shrinkage and lowering of the peat surface being observed, and in dry weather some fissuring occurring. Since it used to be believed that any complete vegetational cover would evaporate about the same amount of water, the phenomenon was puzzling, but advances in hydrology showed that stands of evergreen conifers evaporated considerably larger quantities than ground vegetation. Hence it became realised that even in climates with large annual surpluses of rainfall over potential evaporation, relatively short periods of deficit might induce permanent or semi-permanent changes in the peat towards reduced water holding capacity. Together with root channelling, the peat might thus become progressively more drainable. It has even been suggested that irreversible peat drying is something to be feared, and that the over drainage of peats is possible. In our western climates, however, this hardly seems likely.

Part of the Macaulay Institute's current programme is directed towards physical and chemical changes in peat (especially the mineralisation of nitrogen), and their effects on tree growth.

The post-war programme of drainage experiments has aimed to cover all the main soil types: the deep peats; the numerous gleyed soils on till; and the geological clays. Most has been drainage for new planting, but there has also been work on old woodland sites (especially in the south) and part of the programme has been carried out in growing plantations of various ages, one of the objects here being to find to what age crops will respond in root growth with improved drainage; it being important to know this in view of the large acreages of crops established with inadequate drainage systems.

Drain depths have been taken beyond what is currently considered economic to get a fair picture of the response. Depths of three feet have been aimed at on most sites, and even four feet on certain peats. Where it has not been possible to take out the deepest drains by ploughing, hydraulic excavators have been used. Several of the experiments include alder in mixture with spruce; in one of these (at Halwill, Devon) an early nursing effect of the alder on the spruce was observed well before any below-ground effect could be expected.

A few novel methods have been tried. The ingenious Irish Glenamoy plough, which subsoils a drain in wet peat by extruding a core onto the surface, has been used on several sites. The idea has obvious attractions for giving quick local drainage effects below the planting point, but there has been trouble in keeping the outfalls of the subsoiled channels open at the main drains. Conventional mole draining has been tried in heavy clays and the evidence here has been more favourable, though very few forest sites are likely to be suited to it.

There has been steady developments in ploughs over the period, and one of the recent introductions is a double-throw deep draining plough, which is something of a reversal of the trend. Of the developments in draining technique which have been studied in other countries, the most important (for British conditions) appear to have taken place in Finland. The Finnish Lokomo plough, which is designed to take out very large drains, and is winch-drawn from an anchored tractor, showed some initial promise in trials in this country. An extremely obdurate problem is to mechanise drain cleaning, and so far no ideal solution has been found.

Most of the modern drainage experiments are large in scale and essentially long-term. An interesting exception is an experiment laid down by the Macaulay Institute at Inchnacardoch in which water levels in peat have been automatically controlled, giving very early responses. This, however, is almost a laboratory approach. In order to find out what drainage treatments are doing in terms of soil water, a number of the Commission's experiments have been instrumented to record run-off from the various treatments; one or two are virtually small hydrological enquiries.

The evaluation of drainage effects is likely to prove a complicated business. On many important sites the most direct effect will undoubtedly be the improvement in the stability of the crop, and hence the lengthening of safe rotation. There will also be growth responses, probably quite considerable, but it will be difficult to put a realistic value on late rotation increment on sites with a high natural hazard of windthrow. There is also the question of the longevity of drainage effects, and it may well be that deep draining has a larger effect on the second rotation than the first.

REFERENCES

FOURT, D. F. (1961). The drainage of a heavy clay site. *Rep. Forest Res., Lond.* 1960, 137–151.

HENMAN, D. W. (1963). *Forest drainage.* Res. Brch Pap. For. Commn 26.

HENMAN, D. W. (1965). Some early responses to increased intensity of heathland cultivation. *Rep. Forest Res., Lond.* 1964, 158–165.

MACDONALD, J. A. B. (1950). Afforestation of peat. *Rep. Forest Res., Lond.* 1949, 43–47.

STEWART, G. G. (1959). Preliminary results of experiments in drain deepening in two Border forests. *Rep. Forest Res., Lond.* 1958, 131–137.

TAYLOR, G. G. M. (1970). *Ploughing practice in the Forestry Commission.* Forest Rec., Lond. 73.

YEATMAN, C. W. (1955). *Tree root development on upland heaths.* Bull. For. Commn, Lond. 21.

ZEHETMAYR, J. W. L. (1954). *Experiments in tree planting on peat.* Bull. For. Commn, Lond. 22.

ZEHETMAYR, J. W. L. (1960). *Afforestation of upland heaths.* Bull. For. Commn, Lond. 32.

Chapter 19

REGENERATION AND REHABILITATION

Experimental work on regeneration of woodlands in the post-war period divides roughly into two main activities; the first concerned with the special case of the old broadleaved areas in England, up till 1958; and the second with the regeneration of coniferous plantations generally, which assumed increasing importance from the late fifties to the end of the period. The natural regeneration of native species such as Scots pine and beech had always attracted some attention; in the latter species it was one aspect of the autecological study of the tree conducted by J. M. B. Brown (Chapter 23).

Broadleaved Woodlands

The Commission's broadleaved planting had always been a small proportion of the total; between the wars it did not usually exceed seven per cent. Much of it was restocking old broadleaved woodland sites, and the only considerable broadleaved afforestation was on the Chalk (Part I, Chapter. 4). For the first decade after the second war the Commission continued to plant broadleaved species on sites thought capable of producing hardwood timber of good quality. The silvicultural and economic importance of conifers in mixture had however been appreciated, and pure plantations of broadleaved species became the exception. (Plate 49.)

The Census of Woodlands conducted by J. S. R. Chard (Forestry Commission, 1952–53) and completed in 1949, revealed a very large area of broadleaved woodland in unsatisfactory condition; classified as scrub, devastated woodland, or coppice. It appeared that the total area in need of rehabilitation might not fall short of one million acres. This was regarded by Lord Robinson as a challenge and a subject of high priority for research. The omnibus term "Derelict Woodlands" was coined for the project, and in 1950 A. D. S. Miller was appointed to the Silviculture (South) Section to work on it. As a research project it presented two main difficulties: (i) the material was heterogeneous, and (ii) many of the woodlands were subject to fairly rapid ecological change. There was little pre-war experimental work on the manipulation of coppice and woody regrowth, but during the war, Guillebaud and Sanzen-Baker had started experiments in group and strip planting in mixed coppice at Collingbourne, Wiltshire, and Ffosydd Orles, Monmouthshire.

An early approach in the derelict woodlands project was to lay down demonstration areas for broad silvicultural studies. Two such devastated areas were selected, one at Alton Forest, Hampshire, on loam soils over chalk, and the other on the heavy Oxford Clay at Bernwood Forest, near Oxford. At Alton observations were made for some years on the regrowth of ash and sycamore which had been much damaged by Grey squirrels, and comparisons of a number of conservative treatments were made. Alton provided some useful experience in handling damaged regeneration of ash and sycamore, but was not a particularly rewarding exercise. Bernwood was treated on strictly experimental lines.

The experimental programme was based on a breakdown of the Census information into more detailed classes according to broad soil types and the characteristics of the cover, especially the stage and constituents of the regrowth. Some types, such as old unworked hazel coppice, lent themselves to conventional replicated experiments comparing such treatments as complete clearance and full replanting, planting in groups and strips of various sizes; and planting under more or less uniform cover form singling out the coppice. The experiments laid down at Gardiner Forest, Wiltshire (now Cranborne Chase) on the establishment of beech in old hazel coppice in the fifties were very good examples of this. It was something of a departure that all experiments in the programme were carefully costed. Similar experiments were carried out on other relatively homogeneous cover types, such as *Rhododendron* and Cherry laurel thickets, but here conifers were the principal planting species. It was more difficult to do meaningful experiments in the essentially heterogeneous conditions in many devastated broadleaved woods, but a number of large-scale costed experiments were laid down comparing intensities of effort rather than silvicultural treatments aimed at a common end, e.g., a beech stand. As Ryle (1969) points out, the climate of opinion in the early post-war period favoured conservative low cost treatments such as husbanding all advance growth of useful species in view of the shortage of young/middle-aged hardwoods in the age-class distribution. Miller's experiments wherever appropriate contained such "accept and tend" treatments, and also other variations such as enrichment by limited planting; including the use of large transplants (especially beech) up to 8 ft or more in height, which on the whole were successful. Opinion however swung away from acceptance and enrichment practices, largely because it was realised that investments towards partially stocked broadleaved woods were dubious to say the least, but also because experience was showing the advantages of planting under high cover with the minimum degree of thinning out for survival and early growth. The "dappled shade" prescription of cover associated with O. J. Sangar and R. H. Smith became very general in the southern broad-leaved areas and, as time passed, more and more woodlands became suited to such treatment (Smith, 1961; Stocks, 1961).

In the late forties and early fifties, however, there were large areas of low thicket types of regrowth, usually dominated by coppicing species of little or no value. This sort of condition had been met with in the twenties, and had provided some painful and costly experiences. Machinery appeared a logical approach, and for several years experiments were carried out with several types of machine designed, broadly speaking, either to cut or to uproot, over a range of soils and cover-types.

In retrospect, heavy machinery had perhaps little chance in the broadleaf sector, since the areas (though large in aggregate) were individually small, and no great programme of rehabilitation developed, which would have justified a travelling circus of large machines. The work was, however, useful in showing what machines could usefully accomplish, and equally what they could not do. The anchor chain device, a heavy chain drawn in a loop by two heavy tractors, which had been used in America and Australia (not to mention the ill-fated African ground-nut scheme) for the clearance of scrub, performed cheaply and efficiently under special conditions, such as stiff, scrubby growth on light soils. Such conditions were however rare. Machines such as grubber blades would of course uproot almost anything, but expensively, and the disposal of stumps by burnng was prohibitive. Any means of uprooting coppice or scrub on heavy soils was objectionable, leading to much local puddling. The complete removal of cover on heavy soils whether by machine or hand cutting, resulted in rewetting and the rapid reinvasion of the surface by coarse grasses and *Juncus*. This was accentuated where heavy tractors had been used under moist soil conditions, otherwise it was a phenomenon associated with complete clearance rather than method. Certain machines however could be used under suitable conditions to prepare for planting with the maintenance of cover. The most adaptable instrument proved to be the V-blade cutter, developed by Miller and D. Hampton; a V-shaped dozer blade with sharp lower cutting edge mounted to a heavy crawler tractor. This was capable of cutting stems to 9-inch diameter, and would cut its way through scrub or coppice with comparatively little uprooting; hence it could be used in preparation for strip planting. (Plate 48.)

Active experimentation in the main project closed in 1958, but work has continued on the herbicides (see Chapter 11). Whilst the types of land with which it was concerned (and indeed broadleaved forestry generally) have declined in importance, there were a number of permanent gains. It resulted in a very much better appreciation of the essential features of the various cover types, and the silvicultural alternatives appropriate for each of them. Practice hardened towards one method, that of underplanting lightly thinned high cover, and this of course was encouraged by the gradual development of high cover types. The experiments provided material for a good deal of useful ecological observation on the effects of shade, shelter, and root competition. The main findings of the project, summarised by Wood, Miller and Nimmo (1967) as the Commission's Research and Development Paper No. 51, are not without interest to those concerned in the rehabilitation of broad-leaved woodland for objects other than economic forestry.

Coniferous Forest, Natural Regeneration

In the *Report on forest research*, 1951, J. A. B. Macdonald (1952) reviewed twenty years' work on the regeneration of Scots pine in Caledonian forest remnants. A number of cultural treatments had been applied: heather burning; surface cultivation; manuring etc., and general ecological studies of the incidence of regeneration had also been made. The results had been poor; as Macdonald put it, "unaided natural regeneration is erratic, undependable, and a long drawn out process". Macdonald attributed the difficulties in regenerating the Caledonian remnants to the degradation of the environment following the exploitation of the forest and subsequent introduction of sheep grazing as the dominant form of land usage. There is however little doubt that the tree is at its moist, cool, limit in the north-west of Scotland, which is against frequent and abundant seed production. Its behaviour as a sub-spontaneous species on the heaths in the south-east of England is very different.

No great amount of work was done in the post-war period on the natural regeneration of Scots pine, but in the early sixties seed-fall was assessed by trapping on some areas, showing rather low intensities. Regeneration has been kept under observation, and the Nature Conservancy has taken over the subject in certain of their reserves.

With the exotics, natural seeding remained a matter of academic interest till large-scale plantations of the important species had reached seed-bearing age. James Macdonald (1957) commented that the prospects of the more recent introductions were much better than those of the old exotics. Copious natural regeneration of *Tsuga heterophylla*, *Lawson cypress*, *Thuja plicata*, and to a lesser degree Sitka spruce appeared on a number of private estates where these trees had been planted at an early date. Research interest (naturally enough) centred on Sitka spruce, after it became clear that natural seeding was not rare and ephemeral. It began to appear in windlows, on roadsides, and other disturbed ground, when plantations had reached the age of thirty-five years or thereabouts. It became very noticeable in Neustein's large experiment in the Forest of Ae, Dumfriess-shire, comparing sizes of felling coups for artificial regeneration which was laid down in 1962. (See below.) Seed trapping from 38-year-old Sitka spruce at Glenbranter in Argyll in the comparatively good cone year of 1964 showed seed falls of over a million viable seed per acre, although only some 30 per cent of the crop coned. It was observed that about three-quarters of the seed-fall occurred between February and April, and this pattern was confirmed in the Forest of Ae, when seed trapping of the heavier crop of 1968 gave estimates of seed-fall several times higher. The opportunity was taken to assess the regeneration resulting from these two crop years (1968 for example, gave seedling stockings of about 60,000 per acre) and to study the fate of the seedlings in marked quadrats. As a complementary study, direct sowings of measured amounts of viable seed were made with simple surface treatments to obtain information on the effects of seedbed on germination and survival. As would be expected, the greatest first year losses (in the order of 20 per cent) appeared to be due to drying out of litter, and cultivation had some beneficial effect. The application of phosphate at the Forest of Ae appeared to have none. Studies on natural seedbeds have also been conducted by G. Howells and J. B. Scarratt of Bangor University, working in Gwydyr Forest, Caernarvonshire.

Coniferous Forest—Replanting

The Commission's own plantations were too young for felling during the war. Most of the large area of privately owned conifer forest clear-felled belonged to the easier range of sites; which had not required intensive cultivation or drainage for establishment. On these sites, clear-felling and replanting was the conventional system. There are no problems specific to the "better" site, though weed growth may be more luxurious and the incidence of the fungus *Fomes annosus* higher than on the "poorer" site.

Replanting in the Commission's own forests began to assume some importance and to attract experimental work in the late fifties for several reasons. Windthrow had begun to be significant in spruce forest exceeding forty feet in height on poorly draining soils, especially in the Borders. Considerable areas of early larch plantations were by then putting on very little increment, and there appeared good reason to replace them by higher yielding species. Likewise there were cases where the original choice of species had been faulty.

Another motive for clear-felling, which seemed cogent in the fifties, was to begin the process of normalising the age-class distribution in large forests such as Thetford Chase, where much of the area had been planted over a short space of time with very big annual programmes.

In opening this general project in Scotland in 1958, Edwards and Stewart (1959) put first the question of choice of species in the second rotation, and secondly a group of related questions on the preparation or re-preparation of the site, including the improvement of drainage systems. There were no assumptions about the system for regeneration other than that it was probably undesirable to lose all the benefits of hard-won forest cover. A number of the early experiments were laid down in windblown areas, and it was soon found that deer were a menace in small clearings in extensive forest areas.

Cultivation on stump ground proved difficult, and fortunately on the range of sites so far encountered it has not shown any great benefits. It can be assumed that improved drainage systems will often be required, and whether this should await regeneration or be attempted in standing crops depends on the stability of the forest.

In the Border spruce forest work soon shifted from the special conditions of windblown gaps to larger-scale enquiries on more typical ground. In 1962 Neustein (1965) laid down a large experiment in the Forest of Ae, Dumfriess-shire, in 34-year-old Sitka spruce. This was a shallow rooted crop of low quality class, and the main objects of the experiment were to compare the effects of size of felling coupe on the stability of the crop margins. It was obvious enough at this stage that rotations over much of the Border Forests were going to be governed by windthrow, but it was not known whether there would be any advantage in imposing a less regular structure for the next rotation by preliminary group fellings. The felling coupes ranged from a tenth of an acre to ten acres, and the ground was replanted with Sitka and Norway spruce. Run-of-wind records were made in the different clearings, and observations were made on windthrow on the margins of the crop; there were several other subsidiary enquiries (occurrence of natural regeneration has been mentioned above). The most important observation was that, whilst wind velocity at mid-crown height was least in the smallest clearings, the number of trees thrown *per acre* was greatest here, since the very much increased length of perimeter at risk outweighed the shelter effect. It was also of interest that in this fully stocked crop with no foci of ground vegetation from windblown holes etc., the invasion of grasses was relatively slow, and it appeared likely that planting immediately after felling in such conditions would not require weeding.

In replanting spruces, a short period of check has

been noticed, most markedly in Norway spruce. This was not associated with any obvious deficiency, since the plants responded neither to phosphates nor nitrogen. Root examinations suggested that it was the old story investigated in the twenties by Steven; the spruce habit of regenerating a new superficial root system before beginning to grow. There does not seem any reason to postulate a "replant" problem of the kind encountered in fruit-growing.

The traditional method of disposing of brash in clear-felling was by burning, but this had become increasingly costly. J. S. R. Chard had found it practicable to plant through brash, and claimed that deer browsing was reduced. An alternative to burning was provided by the Wilder-Rainthorpe "chopper", a machine developed from the fodder harvester, and principally used for destroying low scrub. Trials of the machine for disintegrating brash of spruce and pine gave good planting conditions, but costed experiments showed that planting through brash, though expensive in itself, was likely to be a cheaper method than planting with the extra expense of burning or brash chopping.

In the South, the first studies in regeneration were conducted in Scots pine in East Anglia and Japanese larch in Wales. These being relatively stable crops, the obvious approach was to compare various degrees of canopy opening and underplanting with clear-felling. Underplanting of larch (especially) is an old practice, but the current series of experiments were the first to cover a comprehensive range of canopy densities, and a full representation of the most promising successor species. The experiments are still young and full economic judgement must await the complete removal of the overwoods, but some useful indications have been obtained. Optimum canopies have varied somewhat between different parts of the country—they are probably lighter in the wetter districts, for instance—but even tolerant species such as hemlock and Grand fir seem to have done best under fairly light canopies, and the intolerants have usually responded best to full light. At Thetford Chase, where the main adverse factor is the high frequency of growing season frost, it had long been known that cover greatly widened the choice of species available for use. The new series of experiments confirmed the value of cover. Temperature differences between cover and the open were often very large—for instance with 20°F of frost in the open, readings might be as much as 10°F higher under pine overwoods. And as in the pre-war trials a variety of species were satisfactorily established under cover, including Grand fir which has prospects of yielding much more heavily than pines at Thetford Chase. Underplanting is shown on the front cover.

On sites with a history of *Fomes annosus* in the existing crop there has been a particular interest in the susceptibility of successor species, and in view of their reputation for some degree of resistance, *Abies* species have been given special attention. Pathological and entomological aspects of regeneration work are mentioned in Chapter 25.

Experimental work on the regeneration of coniferous forest is still in a preliminary stage, but can be expected to increase rapidly in the next decade as the pre-war plantations reach maturity. It has already shown that it will be feasible to bring in the shade tolerant species such as Western hemlock and *Abies grandis* over considerable areas in succession to pines and larches. In spruce forest, perhaps the biggest question is to what extent natural regeneration of Sitka spruce can be used. Spontaneous regeneration on mineral soils in the west of Scotland is now so common that it seems sensible to look at felling systems and seedbed preparatory measures on a practical scale. On this topic it would be a pity to ignore the possibility of brash burning as a silvicultural measure, since with the progress on fire retardants it is likely that light burns could be undertaken fairly safely and cheaply

REFERENCES

EDWARDS, M. V., and STEWART, G. G. (1959). Silvicultural investigations in the forest: Scotland and north England. *Rep. Forest Res., Lond.* 1958, 46–51.

FORESTRY COMMISSION (1952–53). *Census of Woodlands 1947–49.* In 5 parts. London: HMSO.

MACDONALD, JAMES, et al. (1957). *Exotic forest trees in Great Britain.* Bull. For. Commn, Lond. 30.

MACDONALD, J. A. B. (1952). Natural regeneration of Scots pine woods in the Highlands. *Rep. Forest Res., Lond.* 1951, 26–33.

NEUSTEIN, S. A. (1965). Windthrow on the margins of various sizes of felling area. *Rep. Forest Res., Lond.* 1964, 166–171.

RYLE, G. B. (1969). *Forest service: The first 45 years of the Forestry Commission.* Newton Abbot, Devon: David & Charles.

SMITH, R. H. (1961). The treatment of hardwood scrub. *J. For. Commn* 29 (1960), 17–20.

STOCKS, J. B. (1961). Treatment of degraded hardwood areas for shelterwood restocking. *J. For. Commn* 29 (1960), 20–25.

WOOD, R. F., MILLER, A. D. S., and NIMMO, M. (1967). *Experiments on the rehabilitation of uneconomic broadleaved woodlands.* Res. Dev. Pap. For. Commn, Lond. 51.

Chapter 20

TENDING: WEED CONTROL, BRASHING AND PRUNING

In plantation forestry, tending operations are usually taken to include weeding and such operations as brashing and pruning which precede the main thinnings. Treatments such as fertilisation, drain deepening etc., are thought of as site ameliorations.

Weed Control

By far the greatest share of experimental work on this topic during the period was put into weed control by chemical means. Before the war, no herbicides were used in British woodlands (sodium chlorate had been experimentally applied to *Calluna*), but by 1968 it is estimated that the Commission alone spent about £90,000 on herbicides for routine forest operations.

The development of herbicides in forestry stems directly from research in agriculture and the agro-chemical industry. One or two substances have, however, little use except in connection with woody plants. The Commission's work on herbicides for forest use (as in the parallel nursery investigations) has been concerned with screening new substances and formulations and studying methods of application for particular weed species and stages of growth.

The main programme of work started in 1949 under G. D. Holmes, and was continued by J. R. Aldhous. In the early stages useful guidance was given by Prof. G. E. Blackman's team working on herbicides at Oxford. In 1957 Work Study came into this field, using the methods of operational research on tools and appliances for manual and chemical weeding.

The first object in research on herbicides was to lower the costs of weeding in the rehabilitation of old broadleaved woodlands, in which various coppicing species accounted for much of the expense. Soon afterwards, work extended to various bare land types. Herbicides available to start with included the auxin-type chloro-phenoxyacetic acids (developed principally as selective herbicides for use against broadleaved weeds in cereals), the old herbicide sodium arsenite, and ammonium sulphate, which had been used in America as an arboricide. There was a range of targets for herbicides dependng on the silvicultural methods being pursued; planting in clear-felled broadleaved woodland for instance resulted in a mélange of woody and herbaceous weeds, whereas the maintenance of cover, where available, reduced the vigour of weed growth very markedly. In broadleaved woodlands 2,4,5-trichloro-phenoxyacetic acid (2,4,5-T) (normally in ester formulation) quickly proved a most useful substance, especially for the control of coppice from cut stumps, and also for lightening canopy by killing individual weed trees by the technique of basal bark spraying. In both these treatments it was used as an alternative to the axe or billhook, with considerable economies. Ammonium sulphate (ammate) could be used in very much the same way, but was more expensive and less easy to handle. This substance, however, appeared able to kill almost any woody species, including some (such as ash) which were relatively resistant to 2,4,5-T. It proved especially effective against *Rhododendron ponticum*, which is locally one of the most obdurate and expensive weeds. Latterly, further work on dosage and formulation of 2,4,5-T has also given encouraging results against *Rhododendron*.

Point treatments such as application to individual stools or basal bark treatment require no selectivity, but the treatment of weeds intimately mixed with the crop requires either selectivity or mechanical means for directing the spray. It was found that the chlorophenoxyacetic acids showed some slight degree of selectivity between conifers and broad-leaved weed species, or perhaps it might be said that some of the principal weeds were very much more susceptible. This encouraged the development of overall spraying methods, using emulsifiable esters of 2,4,5-T in water. In this technique, selectivity was accentuated by confining the treatment to periods when conifers were not elongating their shoots, but with the weed species in full leaf, and providing a much larger target than the crop. The use of low volume small droplet sprays reinforced this effect. Such treatments using mist blowers proved very effective in special conditions; for instance where a conifer crop was virtually submerged in a highly susceptible weed species such as gorse or bramble. An obvious extension of this method was to spray coppice or scrub underplanted with conifers from the air, and this was tried on one or two occasions. The dangers of drift with low volume aerial spraying however became very apparent, and the method was not pursued.

One of the early successes with the "growth substances" was the control of *Calluna* using 2,4-D. Here again aerial application had obvious possibilities as a preparatory method before planting heath or moor, but the general objections to large-scale overall spraying (including possible tainting of water supplies) have confined the use of 2,4-D for *Calluna* control to local treatments in checked plantations, a major and critical problem in many upland areas.

In the late fifties further substances became available for trial; the most important being the graminicide dalapon and the bipyridyllium herbicides, which later became familiar under the names diquat and paraquat. These were first tried for the control of weeds in fire traces (Chapter 25), but dalapon with its selective action against grasses and relatively low toxicity to woody subjects was a very obvious weapon against grass swards on sites where these were the main adverse factor. Dalapon proved extremely useful on Chalk downlands, and had it appeared earlier, might well have speeded up the establishment of beech, but planting on this type had virtually ceased by the late fifties. Dalapon however continues to be used wherever grasses are the important weeds, and has considerable advantages in safety and ease of application. Diquat and paraquet (the latter being more the toxic to grasses has proved the more important in forestry) are herbicides with a revolutionary mode of action, since they disrupt the chlorophyll metabolism in the presence of light, but do not affect the older tissues, and are quickly inactivated in the soil. Such materials have great advantages in site preparation since planting can follow very rapidly. In the control of grasses and herbaceous weeds in plantation paraquat has required the development of devices (such as shields carried on the spraying lance) to guard the green tissues and young bark of the plant from the spray. With proper equipment, paraquat advanced very quickly to become the most useful general herbicide against non-woody weeds. While it does not kill vigorous perennials (especially rhizomatous subjects) it does kill a proportion of the grasses, and the general effect is more lasting than mechanical cutting. Under many circumstances it is also cheaper and quicker to spray than to weed with the hook.

Bracken is a weed which is much more important to agriculture than to forestry, since it has little adverse effect in plantations other than physical suppression. In fact, some older foresters have manipulated bracken as shelter. It is however expensive to weed, and the arrival of the bracken specific dicamba in the mid-sixties may be of some minor benefit to forestry. Trials with this substance showed that it is a safe pre-planting treatment for conifers.

Till the late sixties the emphasis had been on herbicides with little or no soil residual effect, though some of the root acting substances had been tried in fire traces. Simazine had found a useful place in nurseries, but its efficient control over seedling weeds was not of much use in the forest, where the problem is usually the deeper rooted perennial herb or grass, often of stoloniferous habit, which may be difficult to eradicate with contact herbicides. Recently, several substances have appeared which offer prospects of useful residual effects on perennial weeds in plantations. Of these, chlorthiamid ("Prefix") has so far attracted most attention. It has the advantage of being applied in granular form, thus saving on water carriage. While it has given very good control of grasses and broadleaved weeds, it has on occasion damaged some tree species.

New compounds in the triazine group, allied to simazine, have also appeared recently, which combine contact toxicity with root action, and these have shown considerable promise. It does however seem that in forestry the target of complete weed control through root action is a rather improbable one, since in the early years after planting there is not much difference to play with in the rooting depth of weeds and crop plants.

Two side-lines with herbicides may be mentioned here. In 1958, some work was done on "chemical pruning" of epicormic shoots of oak and poplar by spraying with 2,4,5-T, following German reports. The method showed some promise but was not followed through as the object was of no great importance. A basic study of the self-pruning habit in certain species might be of interest, as one would imagine the process is auxin controlled.

The increasing importance of hardwood pulp led to investigations in debarking oak, which is one of the species with undesirable bark properties for pulping. Oak coppice and limb wood is expensive to peel by hand, and in the absence of suitable machines some of the "arboricides" were tried. These were applied to the live stem some time before felling. By far the most successful substance for loosening the bark was sodium arsenite, applied to a frill girdle at the base of the stem. A satisfactory technique was worked out (Semple, 1965), but it was decided that the hazards of arsenic were too high to accept.

Research on chemical control has given very useful economies in weeding costs and in shortening the period of weeding on old woodland sites where the vegetation is rather lush. It has had less impact on moorland and heath, where cultural methods are usually adequate for the establishment of the crop without any great expense in weeding. Since the chemists continue to produce new molecules and formulations, one can expect further developments. A higher degree of selectivity between conifers and angiosperm weeds would obviously be an advantage, and so would a herbicide with safe residual effects. But it seems more likely that we shall benefit from higher phytotoxicity in contact translocating chemicals.

The programme has been carried out in a period of increasing anxiety about the hazards of agricultural chemicals to man and the environment. The subject is too big to discuss here, but forest research has worked inside the safety regulations and recommendations of the appropriate government agencies; and in ecological considerations the Commission has collaborated increasingly with the Nature Conservancy.

A certain amount of attention has been given to powered tools for weeding in plantations. These have included self-propelled cutters, either reciprocating or rotary (as on the forage harvester principle), and also back-pack powered cutters like scaled-up hedge trimmers. The powered cutter however is less flexible and pleasant to work with than the modern back-pack spray, and under most circumstances chemical control seems likely to prove more efficient.

Brashing

Weeding in young plantations grades into cleaning—the removal of unwanted woody material—in established plantations. This has attracted no specific study, except as a costed operation in the Derelict Woodlands project. The principal tending treatment is brashing, and the tradition in British conifer plantations has been to "prune" the whole crop to approximately head height; this operation has nothing to do with pruning for timber quality, but serves as a fire control measure and facilitates access for marking thinnings and subsequent extraction. The degree to which plantations should be brashed first received experimental attention in 1948 at the Forest of Ae, but these and later studies conducted by Silviculture in the North did not give very convincing evidence on the economics of the operation. The new Work Study Section investigated brashing as an operation in the late fifties, and later in the period when Work Study was ready to work on harvesting systems, brashing was re-examined in connection with methods of thinning (such as line thinnings for the first intervention) and processes of extraction. This is undoubtedly the correct approach.

Pruning

As a measure for the growing of knot-free timber, pruning has a long history. For certain products, e.g. poplar logs for rotary veneering, it is essential, but otherwise it is a treatment which has to be considered along with others, such as initial spacing, thinning method and rotation length, all of which have a bearing on the size and persistence of knots. Also, knottiness is by no means the only criterion of timber quality. Hence a full economic evaluation of pruning is particularly difficult to achieve, though many aspects of the operation are open to experimental treatment. The Commission's first experiments were carried out by Guillebaud in the early thirties, and besides providing for a comparison of pruned and unpruned timber, enquired into severity (proportion of crown pruned) and intensity (interval between treatments) of pruning. (Plate 50.)

A rather similar series of experiments were carried out in Scotland shortly after the war. A controversial point, especially in the spruces, was the safety or otherwise of green pruning. One view, backed by certain Continental authorities, was that it constituted a pathological risk from frost damage to cambium and the entry of fungi. Cankers have been associated with brashing in Britain (Day, 1951) but this has been a rare occurrence. In 1945 Scott at Princes Risborough dissected knots from spruces green pruned in 1938, and in over 3,000 knots found no instance of disease (Forest Products Research Laboratory, 1952).

In the early experiments it became obvious that too many trees had been pruned, a number of pruned stems having to be removed in thinnings before useful knot free timber had been produced. Likewise, the early experiments raised questions about the effects of pruning on girth (and possibly height) increment, and consequently on the dominance of the pruned tree as against its unpruned neighbours, and these points were looked at more closely in subsequent experiments.

One curious sideline in pruning was the investigation of finger disbudding in pine, carried out in Scotland in the late forties and fifties. This trick (of American origin) removes the lateral buds at the terminal whorl, leaving the axial bud to grow on. On that part of the stem treated the knotty core is virtually non-existent. The problem of course was one of compromise—what amount of this maltreatment would the tree stand, how many whorls to leave etc. Some of the patterns looked extremely odd, and were aptly called "poodle-isation" by J. A. B. Macdonald. It was not found practicable to achieve pruned lengths greater than about 8 ft, and losses of treated stems tended to be rather high. (Henman, 1961).

It had sometimes been claimed by enthusiasts for pruning that the costs could be recouped in the conversion of poles from thinnings, i.e. in snedding and peeling. Timed studies on Douglas fir and Scots pine, however, did not substantiate this.

Certain pre-war pruning experiments were terminated prematurely following wind damage, and these provided an opportunity to look at the patterns of knotty core and clear wood from various pruning treatments. Longitudinal sections through stems pruned on various regimes showed very strikingly how difficult it is to maintain a small knotty core (4 inch has been a common target) with an economical pruning intensity. It is of course self-evident that the rate of diameter increment after pruning is the most important factor in the production of clear wood in a given rotation. The field experimental work is discussed fully by Henman (1963) in Forestry Commission Bulletin 35. The work was quite sufficient to provide practical prescriptions. The evaluation of pruning has been taken further recently, when timber from some of the pre-war pruning experiments has been converted by standard procedure at Princes Risborough and graded according to the Laboratory's *Rules for sawn British softwood*. The main feature after thirty or so years growth following pruning was not the production of significant quantities of clear timber, but the very much higher proportion of sawn material from pruned stems falling in Grades I and II compared with that from the unpruned logs (Forest Products Research Laboratory, 1965). Failing a veneer industry based on home grown logs, this general upgrading seems a more realistic target than the production of clears, and in Scots pine there is a special case for pruning since the dead loose knot is the most important single cause of degrade. However, the whole case for pruning rests on a future for quality timber in the solid, and the trend has been away from this.

REFERENCES

ALDHOUS, J. R. (1962). *Weed control in forest nurseries*. Res. Brch Pap. For. Commn 24.

ALDHOUS, J. R. (1966). *Simazine residues in two forest nursery soils*. Res. Dev. Pap. For. Commn, Lond. 31.

ALDHOUS, J. R. (1967a). *Review of practice and research in weed control in forestry in Great Britain*. Res. Dev. Pap. For. Commn, Lond. 40.

ALDHOUS, J. R. (1967b). *Progress report on Chlorthiamid ("Prefix") in forestry*. Res. Dev. Pap. For. Commn, Lond. 49.

ALDHOUS, J. R. (1969). *Chemical control of weeds in the forest*. Leafl. For. Commn, 51. 2nd ed.

DANNATT, N., and WITTERING, W. O. (1967). *Work study in silvicultural operations with particular reference to weeding*. Res. Dev. Pap. For. Commn, Lond. 58.

DAY, W. R. (1951). *Cambial injuries in a pruned stand of Norway spruce*. Forest Rec., Lond. 4.

FOREST PRODUCTS RESEARCH LABORATORY (1952). *Examination of sawn timber of Sitka and Norway spruce pruned by the Forestry Commission*. Princes Risborough Laboratory, Buckinghamshire (Unpub. report).

FOREST PRODUCTS RESEARCH LABORATORY (1965). *The effect of pruning on the value of home-grown softwoods*. Spec. Rep. Forest Prod. Res. 22.

HENMAN, D. W. (1961). Pruning of conifers by disbudding. *Rep. Forest Res., Lond.* 1960, 166–172.

HENMAN, D. W. (1963). *Pruning conifers for the production of quality timber*. Bull. For. Commn, Lond. 35.

HOLMES, G. D. (1957a). *Experiments on the chemical control of Rhododendron ponticum*. Forest Rec., Lond. 34.

HOLMES, G. D. (1957b). *Notes on forest use of chemicals for control of unwanted trees and woody growth*. Res. Brch Pap. For. Commn. 21.

SEMPLE, R. M. G. (1965). Further trials of chemicals for de-barking hardwood pulpwood. *Rep. Forest Res., Lond.* 1964, 195–199.

Chapter 21

CROP COMPOSITION, TREE STABILITY AND RELATED TOPICS

Crop composition is primarily a character of the silvicultural system governing felling and regeneration. In fact extremely little work has been done in Britain directly on the silviculture of a particular species in the classic sense. Clear felling with artificial regeneration is the only widely practised system, though locally there have been silvicultural practices depending on natural regeneration with beech and Scots pine, and in recent times certain private estates have used natural regeneration of the more recently introduced conifers in some approach to *Dauerwald*. But so far as Commission research is concerned, most enquiries into differences in crop composition have arisen from different methods of establishing the crop, and have not been concerned with systems of regeneration. Hence the lines of work mentioned in this Chapter, though all concerned in one way or another with crop composition, do not present a coherent picture.

Nursing Mixtures
Nursing is one of the old lines of work, and remained active for a number of years after the war. Its aim was to promote favourable ecological conditions for a major species by means of an easily established pioneer crop; either a tree or a shrub. Crop species and nurse might be planted together, as in the fashionable pine/spruce mixtures of the forties, or the crop tree might be introduced into the nursing medium some years after establishment. As mentioned in Part I, Chapter 4, remarkable successes were achieved with common broom (*Sarothamnus scoparius*) as a nurse crop on heaths for *Calluna*-sensitive trees such as spruces, *Tsuga* and Douglas fir. Broom continued to be used in establishment experiments on different sites for several years after the war, and the method was considerably improved. It never entered into general practice, and fell out of consideration because of its expense, and the contemporary developments in cultivation and establishment methods. The use of tree species as nurses however did not involve expense on a non-productive element of the crop, and interest in these lasted longer.

Most of the early nursing mixtures (experimental or otherwise) were laid down in intimate arrangements, row by row or tree by tree. These were difficult to handle—a nurse capable of suppressing ground vegetation quickly was equally likely to suppress the nursed species. However, Guillebaud had recorded improved height growth in spruce in such intimate mixtures with pine before the war. In some cases pine mixtures were effective in establish-

ing frost sensitive species such as Douglar fir. But the best evidence on nursing effects was obtained fortuitously, from observations of edge effects in experimental plots established for other purposes. At Wykeham, Yorkshire, it was observed that slow-growing spruce and other species were stimulated by proximity to pure plots of pine and larch which had closed canopy and suppressed the heather. It was found that spruce roots had invaded these plots and were exploiting the organic layers. This led directly to the planting of nursing mixtures with bands of sufficient width (usually three rows) of the nurse species to provide a zone of heather suppression. For a short time such mixtures (usually of pines and spruce) were commonly planted on moorland and upland heath. However, this line was overtaken by the developments in ploughing, manuring, and chemical control of *Calluna*, all of which made it easier to establish the major crop pure; with adequate cultural methods nursing only seems to give an extra benefit under exceptional, marginal, conditions. Certainly with Lodgepole pine and Sitka spruce which has been the commonest mixture, the tendency has been for either the spruce to dominate on the best sites or for the pine to do so on poorer sites, leaving few circumstances where the pine has clearly been beneficial to the spruce. Hence nursing mixtures for the short-term effect have not proved an important line.

Robinson's ideas of "accelerated succession" suggested the introduction of successor species into holes cut in thicket stage crops of pines. Generally speaking, such interventions in vigorously growing young stands proved unmanageable; though Western hemlock was successfully established this way at Teindland and other heathland sites where a representation of the species was the only criterion.

Long-term Mixtures
These are distinguished from nursing mixtures by the object, which is to grow them to maturity as such. In practice the distinction is not always clear. The topic is one on which there has always been much argument—it is apt to shade off into wider considerations such as irregularity of crop structure. Very little hard evidence on the benefits of mixtures was available in the early fifties though many had been planted. In 1953 M. V. Laurie prepared an experimental plan to investigate simple two-species mixtures, and over the next few years a considerable number of these experiments were laid down. The intention was to compare each species pure with the mixture between them in respect of the crop

characteristics (production being the most important question), and the effects of the species pure and in mixture on the site. These are essentially long-term experiments, and include most of the conventional or plausible combinations of species. There was no expectation of increased production over one rotation (whatever other advantages might accrue) where a high yielding tree was being "diluted" by another of normally lower increment, but it seemed possible that mixtures such as Douglas fir or Sitka spruce and Western hemlock might make a more efficient use of the supply factors—light, water, fertility—than pure crops of these species. However, where the most productive species on any site is itself a shade tolerant tree such as *Abies grandis* or Western hemlock it seems most unlikely that admixture will increase the yield over a single rotation.

The effects of species on the site and the whole question of fertility maintenance is a very wide subject, and something will be said in Chapter 23 about fundamental work conducted by Oxford, the Macaulay Institute and the Nature Conservancy. The Commission's two-species mixture experiments included broadleaved trees with a reputation as soil improvers, and were laid out to facilitate soil studies. One experiment was established in collaboration with the Nature Conservancy specifically for this purpose.

As might be expected a number of the experiments ran into difficulties, especially those containing *Tsuga*, which is extremely difficult to establish on bare sites. In mixture with Sitka spruce, some early nursing effects were observed. At the present time, mixtures are rarely planted, but this set of experiments may have some bearing on conditions in the second and later rotations.

Spacement in Plantation (Plate 53.)

A few experiments had been established in the early nineteen-twenties, but the main effort dates from 1935 when James Macdonald drew up a plan for a standard type of spacing comparison. Over 140 experiments to this plan were laid down by the "Divisions". They were simple unreplicated trials of initial planting spacings ranging (usually) from three to eight feet. Many of the experiments suffered from labour difficulties during the war. In 1949 the series was assessed by Research and yielded useful information on rates of closure of canopy and other early crop characteristics; some of the more forward experiments also gave evidence on branch dimensions and stem taper in relation to spacing. In 1953 a new plan was prepared for the future conduct of the best of the spacing experiments, introducing contrasting thinning treatments; some experiments being thinned on conservative lines to attain a common stand density, whilst in others the different rates of

diameter increment due to spacing were maintained or accentuated by thinning the wider spacings more heavily than the closer ones. The latter treatment was aimed specifically at producing timbers of very contrasting ring widths for study at Princes Risborough.

Like some other lines of work, spacing at planting was somewhat uneasily placed between Silviculture and Mensuration, and the topic has perhaps not had the attention it deserves. Practice was forced off the closer spacings some time ago by rising costs, but it is doubtful whether the widest spacings adopted today can be supported by the evidence. However, the spacing experiments have paid dividends in a variety of directions. They helped to fill a gap in the mensurational evidence on stocking and yield, and certain of the experiments have lately been used in fundamental studies of the stand as a productive mechanism. They were also useful in work on stability as they provided crops of different densities and stem characters on the same site. And as intended, they have given useful material for the forest products specialist. Brazier (1967–70) especially has made good use of spacing experiments in studying width of ring and early and late wood development.

Uneven-aged Stands

The commonest and simplest form in British forestry is the result of underplanting a light demander (usually larch) with a shade tolerant tree. The practice of course is old and well established before the Commission's time. Most early investigations into such crops have been concerned with the yield of two-storeyed stands; for example, the well-known series of underplanted larch sample plots at Dymock, Gloucestershire, and Haldon, Devon (Earl, 1951). That the method succeeds in increasing yield over the larch rotation hardly requires proof, and it is also an excellent method of establishing frost-sensitive species; but as yet there has been no full economic analysis of the practice. The recent series of experiments mentioned in Chapter 10 on the topic of replacement of larch crops should however give further quantitative evidence; it is certainly high time that so old a practice should receive a proper appreciation. Most underplanting today is aimed at a change of species rather than at long-term two-storeyed crops. (Plate 51.)

Stability of Trees and Crops

This topic became a research project in 1960; the instability of spruce crops on ill-drained sites had however been a source of anxiety for some time before this, and it had been expected that some of the early drainage experiments would give evidence on rooting depth and resistance to wind throw. A. I. Fraser adopted three lines of investigation: (i) the incidence of windthrow in relation to gales and locality factors, (ii) the forces acting on the individual tree, and (iii) aerodynamic studies on the crop. A national reporting system yielded useful statistics of the occurrence of sizeable windthrows in relation to gales of known strengths and locality factors. Gales gusting to 60 mph occur somewhere in Britain every year with damage increasing in aggregate as more crops reach susceptible heights (40 ft or so). The behaviour of crops in such normal gales is of more research interest than the consequence of hurricanes such as those of January 1953 in northern, and January 1968 in western and central Scotland; for very little can be done about these.

The forces on the individual tree were looked at first by measuring the wind-loading by dynamometers on guyed stems. Resistance of stems to throw was studied by the technique of measuring the pull required to displace them by a winch (Fraser and Gardiner, 1967). This method, developed by Fraser, owes something to J. A. B. Maconald and also to T. R. Peace, who had been interested in "winching-out" as a quick method of excavation in pathological root studies. From the first field studies of wind-loading, Fraser turned logically to the wind-tunnel, and was fortunate in securing the interest and expertise of the Royal Aeronautical Establishment in carrying out studies on small tree crowns in the 24-ft wind tunnel at Farnborough. These tests, which measured the stresses on the butt from wind speeds in the range 17–50 knots were most informative. It was observed that drag in this range was directly proportional to wind speed, not to the square of the velocity as in solid structures; the distinction being due to the tree's ability to stream-line itself progressively. Also, it appeared that the weight of tress explained most of the differences between species and shapes of crown in the drag/wind speed relationships (Fraser, 1963). This work suggested the construction of small model trees with the correct drag characteristics for a given wind velocity, and an electrical device to measure loading, to study the behaviour of "crops" in smaller wind tunnels; and in 1963 Fraser collaborated with D. E. Walshe of the National Physical Laboratory in some tests of crop arrangements in the 9 ft × 7 ft open-jet wind tunnel at Teddington (Walshe and Fraser, 1963). The approach has too many limitations to be of major importance, but it certainly suggested that any increase in roughness was likely to be accomplished by greatly increased forces on the crowns of individual trees. (Plate 52.)

Tree-pulling studies yielded valuable information on a number of points for a comparatively small effort. Firstly they confirmed Day (1949) in showing

the great sensitivity of the rooting system of Sitka spruce to compacted or anerobic soil conditions. In good soil conditions, Sitka spruce has a deep and searching root system. Under Border conditions, drainage had deepened the rooting systems, and where this had enabled the roots to penetrate below the peats into the till there was some gain in stability. This was further evidence to encourage the adoption of deeper drainage systems. Tree-pulling studies in spacing experiments showed that closely spaced crops rooted deeper than widely spaced ones; a most interesting observation confirmed by tree-pulling in "isolation" experiments (an extreme form of crown thinning) where the favoured trees, though much larger than those in the crop matrix, were comparatively shallow rooted. It was assumed that this effect on moist sites was hydrological, denser crops drying out the soil to greater depths at an early stage of growth. It will be of interest to see whether this effect is accentuated in extremely dense crops resulting from natural regeneration.

Carried out over a range of soil types, and in relation to the evidence on wind-loading on tree crowns, such studies led to estimates of the susceptibility of crops to windthrow on an actuarial basis, and this was taken up by D. G. Pyatt, then of the Management Division. Such estimates are of obvious importance in the preparation of working plans for forests where the hazards are known to be high because of soil and topographic factors, and this is the most practical outcome of the project. However, there remains a good deal to be learned about windrun over forests, and in 1967 a large-scale experiment was established at Redesdale, Northumberland, using an automatic data logger to record anemometer readings from various levels on towers set up in a forest crop and on open ground outside it. The intention in this experiment is to compare thinned and unthinned crops in respect of the characteristics of air movements in and over the canopies, but so far only the differences in aerodynamic characteristics between conditions in the open and the undisturbed forest have been studied (Fraser, 1968).

A good deal of other observational work has been done on stability questions. In the sporadic (single tree or small group) windthrows in Border forests, surveys by Neustein showed that trees adjacent to freshly cleaned drains were usually the first to go. It is unlikely that the problem of cleaning neglected drain channels in spruce forests is wholly soluble; there is bound to be some loss of stability. Numerous observations following gales have led to general conclusions on the influence of topographic features; the great vulnerability of forward slopes or slopes diagonal to the wind, funnel effects in valleys etc.

Nothing in the investigation on stability has

offered much encouragement to the belief that ir-regular crops are inherently more stable than homo-geneous even-aged stands, which does not necessarily imply that diversification has no place in reducing the hazard of large-scale windthrow.

REFERENCES

BRAZIER, J. D. (1967–70). Timber improvement. 1. A study of the variation in wood characteristics in young Sitka spruce. *Forestry* **40** (2), 117–128; 2. The effect of vigour on young growth Sitka spruce. *Forestry* **43** (2), 135–150.

DAY, W. R. (1949). The soil conditions which determine wind-throw in forests. *Forestry* **23** (2), 90–95.

EARL, D. E. (1951). *A short note on the underplanted European larch plots at Dymock, Gloucestershire, and Haldon, Devon.* Unpub. MS, Forestry Commission, Planning and Economics Branch.

FRASER, A. I. (1963). Wind-tunnel studies of the forces acting on the crowns of small trees. *Rep. Forest Res., Lond.* 1962, 178–183.

FRASER, A. I. (1968). Tree stability: Aerodynamic studies. *Rep. Forest Res., Lond.,* 83–84.

FRASER, A. I., and GARDINER, J. B. H. (1967). *Rooting and stability in Sitka spruce.* Bull. For. Commn, Lond. 40.

WALSHE, D. E., and FRASER, A. I. (1963). *Wind-tunnel tests on a model forest.* Aero Rep. 1078, Natn. Phys. Lab., Teddington.

Chapter 22

PROVENANCE AND TREE BREEDING

Although the objectives in these two fields of work are the same, it was found convenient to leave provenance—i.e. studies on the broad geographic and ecotypic variations in exotics—with the Silvicultural sections, leaving the new Genetics Section to concentrate on more intensive work on variation in home forests and on the process of tree breeding. This served well enough under British circumstances, since the fund of variation from past introductions and provenance experiments was very great. It is not an arrangement one would expect to see perpetuated, as there is no point in carrying on provenance work in exotics after it is clear what the major seed source variations are; when further improvement is a matter of developing home seed sources and tree breeding.

Provenance

By the end of the war sufficient experiments had been laid down to give a reasonable idea of the *scale* of provenance differences likely to be experienced in the major species, and in one or two instances (especially Scots pine) a fair picture of the *pattern* of variation had emerged. It was clear that the British climate (except at the extremes) was not a highly critical screen for provenance: a number of the important exotics could be established from seed sources in large parts of their natural range, and the results of misfits in climate, latitude, and elevation, had sometimes shown themselves belatedly in pathological symptoms in thicket stage crops.

In the early post-war period there was a great deal of assessment to be done in the established experiments, and this had priority over new work for some time. At this stage much attention was given to assessments of form of stem, branch habit, health etc. in addition to simple measurements of growth rate. Also the design and general conduct of provenance experiments was studied; the older experiments had pointed to various difficulties in carrying out such long-term experiments. This was a topic on which M. V. Edwards did important work. Later in the period advances in experimental statistics made available new designs, which were introduced by J. N. R. Jeffers; these were particularly important when the numbers of provenances represented in experiments began to rise.

One aspect of provenance work remained controversial over most of the period—the nature of the representative collection. Broadly speaking these may be collections from relatively few trees at a described locality, or much larger collections (usually commercial) from a region or zone. In the latter case what is sown for the experiment is a small sample.

Some experiments have included both types of collection. Commercial collections are not favoured for experimental work, but it should be pointed out that where the collecting zones are defined and the methods regulated, such collections do sample the range of a species in a meaningful way and the results are reproducible. For the limited objective of finding the best zones for large-scale importations, samples from commercial collections have served very well, and much of what we know of the provenance of species such as Douglas fir and Sitka spruce is based on comparisons of seed from the collecting zones of the long-established American firm of Manning. Collections of this sort (commercial or otherwise) fail to show the more intimate patterns of variation, and there may also be bias towards poorer genotypes; perhaps especially towards precocious and branchy forms of tree. Also, there may be no organisation for collecting seed from certain localities, simply because there has been no market for it. The best way of conducting collections for provenance studies is to send out a team and conduct a proper sampling both between zones and inside them. IUFRO, with assistance from the Forestry Commission, have sampled the distribution of Lodgepole pine, Douglas fir and Sitka spruce over the past five years, and the resulting provenance experiments have been arranged on an international basis.

Progress in provenance work has been reported regularly and at length in the *Report on forest research*, and comment here will be confined as far as possible to changing approaches and objectives. The first major post-war activity was concerned with European larch. Provenance work before the war was aimed primarily at dieback of larch, a symptom which usually appears at the thicket stage and is associated with the fungus *Trichoscyphella wilkommii*, and perhaps also with spring frost. Early experiments compared larch of various Scottish sources with Alpine larch (as imported, unfortunately, in quantity) and Sudeten larch, though this was often not the true article, but seed from plantations in that region. Scottish strains were variable, but the best showed some superiority over high alpine provenances in respect of resistance to dieback.

The first big experiment after the war was the international European larch provenance experiment planted in 1946/47 from seed collected during the war years under the auspices of IUFRO. This contained 44 seed lots, and incidentally strained our capabilities for experimental design. Though not fully comprehensive (it did not for instance include representations from the Polish outliers), it has provided a mass of information on locality, elevation etc. With other evidence it has confirmed the superiority of Sudeten larch over the old Scottish strain, which is thought to be of alpine origin. In the fifties Polish larch was introduced to experiments, and impressed greatly by its vigour, though form left something to be desired. European larch however has declined in status over recent years and for the diminished acreage now planted the development of home seed sources is the obvious answer.

Part of the decline in planting of European larch may be attributed to competition with the more accommodating Japanese larch, though the larches as a group have been somewhat out of favour. Japanese larch as imported commercially from Nagano Prefecture in Honshu Island has never presented any serious problem suggesting provenance remedies, apart from the hope of improving stem straightness. Work would probably not have been attempted for British needs, but in 1959 the opportunity was presented to lay down a respectable experiment due to the kindness of Dr Langner of Schmalenbeck, who had recently collaborated with Dr Iwakawa in a comprehensive collection. The indications of this experiment are that we have not been far out in our sources of introductions, and differences are not dramatic.

Norway spruce, the oldest of the exotic conifers, had given little trouble in cultivation, and there were no specific aims in early provenance work. Variations in date of flushing were well known at the individual tree level, and it was realised that a provenance with later than average flushing date would be useful. It has a rather unsatisfactory provenance history. Broad collections were included in experiments laid down in the late thirties and war years, mainly in the Border forests; but inexact details of seed origin, irregularities in planting stock, exceptional frosts and one disastrous fire vitiated much of the evidence. This has proved the more unfortunate, as the weight of evidence is strongly in favour of provenances from south-eastern Europe; a region which has been neglected in seed importation. Recent comprehensive experiments are beginning to confirm this belief.

Scots pine had the best and most comprehensive experiments laid down before the war. By the late forties it was possible to dismiss all high latitude European provenances, being much too slow in growth, and the southern fringe of the species had presented difficulties in establishment, and often very coarse growth. Whilst certain provenances from our own latitudes in Europe gave very good results (especially in south-east England), there was little suggestion that there was much to be gained from foreign sources, particularly for the main pine growing areas in East Scotland. There was little point

in pursuing provenance of Scots pine to increase the production of pine timber inside the range of Corsican pine, and for the main pine growing districts in eastern Scotland the best compromise between the fine form of northern provenances and the faster growth of southern strains seemed to lie in the better Scottish provenances, not all of which are purely Caledonian in origin. One interesting sideline in Scots pine provenance work was the search for pines resistant to winter blasting on the western seaboard. It might have been expected that this character would be found amongst the most western Caledonian remnants, and by the early sixties this was established, but not at a very useful level in view of the much higher resistance of coastal provenances of Lodgepole pine. Mention should be made of the late Capt. Brander Dunbar's curious experiment at Pitgaveny Estate, north-east Scotland, though perhaps it is nearer breeding than provenance. He made a practice of sieving his seed, and sowing only the larger fraction, claiming (quite correctly) that he obtained larger seedlings thereby. Having done this over more than one generation, he convinced himself that he had gone some way to producing a "giant" strain, and certainly he appeared to have made an effective selection for cone and seed size. Efforts to persuade him that he was probably not selecting for anything else were not successful! Indeed plantations of his "giant" strain seem to have held their nursery vigour longer than might be expected. Seed size is a troublesome element in provenance work, and it has become common practice to look for regressions on seed size in nursery and early field measurements.

The other important Old-world pine, *Pinus nigra*, is, like European larch and Norway spruce, a long-established exotic. A tree of discontinuous range, its geographical variants are usually accorded sub-specific rank. The attributes of two of these, Corsican pine (*P. nigra* var. *maritima*) and Austrian pine (*P. nigra* var. *austriaca*) have long been known in Britain. The latter has valuable properties, being particularly resistant to exposure, but there was never any doubt that Corsican pine was the variety required for general planting, and seed obtained from Corsica normally gave excellent results, especially on sands and gravels in the south and east of England. In the late twenties some dubious material was imported from Corsica and planted on some scale in East Anglia. Straight-needled and bushy in habit, this appears to be a high-elevation form of the tree, but it has never been satisfactorily identified and has been (rather unkindly) known as "Ursuline pine" after the collecting agent concerned. However, Corsican pine as such threw up no problems warranting provenance enquiries till extensive browning and, more serious, dieback associated with the fungus *Brunchorstia*, appeared on marginal sites. It was fairly clear that the tree had been stretched beyond its limits to cool, moist environments and over-elevated sites. There was however a special case for Corsican pine (or more generally speaking Black pine) in those parts of the Pennines subject to industrial pollution, and in the late fifties fairly comprehensive experiments were laid down. These concentrated on seed sources from Europe where the tree had been long cultivated well to the north of its natural range, such as the famous plantations at Koekelare in Belgium and Les Barres in France, both thought to be of Calabrian rather than Corsican origin. Calabrian pine had in fact appeared in limited experiments in Wales somewhat earlier, and had promised well at fairly high elevations; it had also shown the useful property of high survival on planting. Whilst some of these provenances are promising, there does not seem to be room for any great extension of the range of Corsican pine under present planting policy.

The last of the European species to attract serious provenance studies is the Silver fir, *Abies alba*. This is a special case as the tree was written-off in the early days of the Commission because of the serious damage caused by the adelgid *Adelges* in young plantations, though apart from this, the large dimensions attained by many individual trees (often in exposed positions) suggest that the European Silver fir is very much at home. There is also a special interest in *Abies* species in view of the reputed resistance of some of them to the fungus *Fomes annosus*. In the very active post-war period of provenance studies it was thought worthwhile to examine the geographical variation in *Abies alba* with resistance to *Adelges* as the main criterion, and in the mid-sixties a collection of 36 provenances was established on several sites. It is of course too early for any indications of value.

Turning to the American trees; in Douglas fir the differences in behaviour of the major geographic varieties were known before the Commission's time, and comparisons of the type "green" Douglas and "blue" Rocky Mountain forms (var. *glauca*) were to be seen in some of the old Forest Gardens laid down in the early years of the century. In particular, the susceptibility of southern mountain forms to the needle cast disease *Rhabdocline* was recognised. Not so much was known about the so-called intermediate or Fraser River Douglas fir (var. *caesia*) typical of the continental distribution of the tree in the northern half of its range. Douglas' own introduction (probably from Fort Vancouver, near Portland, Oregon) is now thought to have been a very fortunate seed source, and some excellent plantations have descended from it. It seems likely too that the large importations from the Lower Fraser Valley, arranged through the Canadian Government in the early

twenties, were also excellent material. Some later imports from rather dry, warm climates in Washington south of the Puget Sound were less fortunate, and may have contributed something to the failures of Douglas fir when pushed too far into moist western conditions. In eastern England, where climate and some at least of the soils favoured Douglas, one limiting factor was the susceptibility of the species to late spring frost. That this might have a provenance remedy was suspected by the behaviour of one very successful plantation in the species trials at Olleys Farm, Thetford Chase, which was markedly late flushing in habit, though unfortunately its origin was unknown.

No very good experiments had been laid down before the fifties, by which time the obvious target appeared to be a study of the main coastal distribution of the tree. This was arranged by sampling the collecting zones in Washington and Oregon of the firm of Manning, which had been drawn up in consultation with Leo Isaac, a leading American silvicultural authority on Douglas fir. The experiment was planted at a number of sites in Britain chosen for a wide range of climate including sites likely to present a hazard of spring frost. Whilst it is obvious that Douglas does not present a simple clinal pattern of variation, this experiment did indicate some very useful generalities. Provenances from the coastal limits of the tree in Washington (one might say the overlap with Sitka spruce) have on the whole shown the greatest promise, and have combined vigour of growth with a tendency to flush late. This on one or two occasions has been sufficient to prevent serious damage when many other provenances have suffered badly from frost.

Some interest in Fraser River Douglas fir (var. *caesia*) has been maintained, since in Scottish experiments this form has grown quite well, and has appeared markedly less prone to winter cold damage on exposed upland sites. It has however shown itself susceptible to *Rhabdocline*. Till recently there has been one serious gap in our provenance coverage in that we have not had material from the northern coastal limit of Douglas fir in Vancouver Island. This has however now been remedied. But it seems likely that for the useful British range of this fine tree we are in the fortunate position to be able to use seed from regions not far from the optimum of the species, with a bias towards the coastal fringe.

Sitka spruce has had a more constant seed source than any other American species; the Queen Charlotte Islands having provided the main supply, and coastal Washington a much smaller proportion, though the pre-Commission plantations at Inverliever and elsewhere probably originated from the latter region. The first experiment at Radnor in Wales, planted in 1927, had suggested a simple pattern of variation according to the latitude of the seed source, with marked increase in vigour from Queen Charlotte Islands through Washington to Oregon; but serious frost losses were sustained with seed from as far south as the northern tip of California. The potential gain in production in taking seed from Washington (let alone Oregon) was however considerable, and the main object of subsequent provenance work in Sitka was to assess the risk from more southerly seed sources. One very obvious hazard had been noted before comprehensive experiments had been laid down: seedlings of Washington origin were apt to grow late into the autumn and might suffer from early frost before hardening-off buds. The first comprehensive experiment in Sitka spruce was sown in 1958, and yielded most valuable information in the nursery stage. The latitudinal clinal pattern of growth rate was clearly demonstrated, and phenological observations showed that this was a matter of the length of the growing period among provenances, rather than date of flushing or the actual speed of elongation whilst growing. It later became clear that susceptibility to *spring* frost damage on the planting site was little related to provenance, in fact faster southern provenances might occasionally escape from the zone of shallow inversions before slower northern ones. But it also became apparent, especially after the severe season of 1962/3, that extreme winter cold might do serious damage to provenances south of the Queen Charlotte Islands, and that there was a considerable risk in using these on elevated and highly exposed sites. Even Queen Charlotte Island material has sometimes been damaged by winter cold, and a special line of enquiry has been to discover whether there is any advantage in provenances from south-east Alaska for the severest climatic conditions. However, comparisons of Queen Charlotte Island and Alaskan spruce over a wide range of sites have suggested that the area over which the latter is likely to have an advantage is very small—probably insignificant.

Sitka spruce is one of the exotics which is now starting to bear large quantities of seed of good quality in the old stands, and we should be independent of imported seed in the near future.

Lodgepole pine has presented the most complex problems of the American species. It has a very great range, but its distribution is discontinuous. Like Douglas fir, the main geographic variants are often accorded sub-specific rank. Its status has changed very greatly over the post-war period. It has found its place as a major plantation species since the war, and the provenance side of this story is almost entirely due to the appreciation of the value of the coastal form of the tree, which is usually regarded as the type species, *Pinus contorta* (Loudon). This has been more a matter of field observation than of experimentation, since the older experiments were not sited in what are now the important environments for Lodgepole pine, and in any case they tended to concentrate on the continental races. These, generally speaking, have far superior stem form to coastal material, but continental Lodgepole pine at its best competes with Scots pine, without any very marked advantages, though it is certainly more productive on the poorest soils. It has however no place at all in the wet and exposed western regions. Nearly all post-war provenance work has concentrated on Lodgepole pine of coastal or intermediate characteristics. (Plates 22 to 26.)

It has been most useful to us that the Irish planted coastal Lodgepole pine pure on an extensive scale well before we did. Their plantations, after one or two thinnings, did a great deal to remove the fear that such provenances were worthless for timber production. The great incentive to further work on Lodgepole pine was of course the fact that considerable areas of poor peatlands were being acquired in Scotland on which Lodgepole pine was the only species which would grow with certainty. For a number of years after the war, a common seed source of Lodgepole pine was Lulu Island, an impoverished tract at the mouth of the Fraser River, in British Columbia. This proved to be a rather precocious strain, lacking in the vigour and resistance to exposure of the material from the Washington coast, which has become the standard seed source.

There are two main directions in which we should like to improve on Washington coastal material. Firstly, there is still the question of stem form, especially on peats, where Lodgepole pine tends to rock in its early life and exhibit basal bowing as it stabilises. The provenance remedy for this may be to accept slower initial growth, and the form of the northern coastal provenances from the Queen Charlotte Islands and south-east Alaska is promising.

Another line is to take seed from places where there is a chance of gene exchange between coastal and continental populations, hoping for intermediate characters with some of the advantages of both. The Skeena River in northern British Columbia appears to be one of the few places where such intermediate forms might develop. For many years it was difficult to obtain seed from these northern coastal areas or from the Skeena River, and in 1967 it was decided to send out a small team (H. A. Maxwell and J. R. Aldhous) to explore the possibilities and arrange for collections. In addition to laying down provenance experiments, seed from these regions has been allocated for raising special seed stands, since it is apparent that coning in the northern part of its range is irregular.

Several other north-west American trees have

attracted provenance studies in recent years, including the potential major species Western hemlock (*Tsuga heterophylla*) and *Abies grandis*. The former, which is difficult to establish on bare land, shows some interesting interactions between provenance and locality factors, in that southern provenances have grown faster under canopy in the south of the country, whereas northern provenances have done better in exposed conditions in Scotland.

Tree Breeding (Plates 55, 58, 59.)

J. D. Matthews, now Professor of Forestry at Aberdeen, in organising the programme of the new Genetics Section formed in 1947, had the example of such pioneers in the field as Syrach Larsen in Denmark and Lindquist in Sweden. The programme was based in the first instance on mass selection methods. In the short term this consisted of the selection and registration of superior stands for use as seed sources; in the mid term the selection and vegetative propagation of superior trees (plus trees) and the subsequent establishment of clonal seed orchards based on mixtures of those vegetatively propagated clones; in the longer-term progeny testing the selected clones and the establishment of second-generation clonal orchards based on proven good clones. In addition, plans were made for some hybridisation and recurrent selection work—all of which leads to the production of new synthetic varieties. The programme was directed at all times towards practical ends, i.e. the necessary development work to reap improved seed at any level was a component part of the work.

The first list of trees selected for concentrated attention included European larch, Japanese larch, Scots pine, Douglas fir, beech and Sitka spruce. Larch now appears a curious choice, but with the considerable knowledge of its variation and the existence of an important hybrid it was at the time an obvious subject.

The primary task was the survey of the potential breeding material. This process sorted out and marked down superior individual phenotypes—the basic breeding material—and at the same time stands containing high proportions of stems of desirable characteristics were identified. This latter was the basis of the register of seed sources. The characters governing selection naturally varied between species, but in all cases vigour, stem form and health were the primary criteria. Generally, the other characters were those which might be expected to influence timber quality such as branch habit and angle, etc; but in the early stages little was known about the heritability of some of the important variations.

The processes of breeding required much work on the development of suitable techniques especially in vegetative propagation; this being the practical means of collecting numerous genotypes together at the centre. Most of the trees worked on in the Geneticist's programme (poplars and elms remained a sideline) were difficult to propagate by cuttings, and grafting remained the principal method over the period. The exceptions amongst the conifers were Western red cedar (*Thuja plicata*) and the hybrid cypress x *Cupressocyparis leylandii*. Grafting itself raised numerous problems. The mechanical side of it—the adaptation of known horticultural techniques to the various species—was mastered early on, and with potted stocks under glass success rates of 60 per cent or more soon became commonplace. Grafting in the field to established stocks gave good "takes" of Scots pine, Douglas fir and larch, but was handicapped by a restricted grafting period in April. The "take" on larch was improved by enclosing the union with a perforated polythene bag. A good deal of attention was paid to stock-scion interaction, including choice of species for stocks; the conifers permitting a surprisingly wide range of combination. Compared with fruit growing practice, no important gains came from this line; the virtual impossibility of working up clonal stocks for many conifers was a great restriction. Grafting incompatibilities between individuals within the same species were often observed, but except for Douglas fir were not common enough to be a serious handicap. A more important trouble, for species other than pines, was the failure of scions to develop apical dominance quickly (a well-known phenomenon in grafted ornamental conifers). The practice of taking scions from vigorous shoots high in the crown occasionally succeeded, but plagiotropic development of scions was often troublesome in Douglas fir.

The alternative method of vegetative reproduction —by cuttings—made uneven progress. The introduction of the mist technique in 1956/7 was a big jump forward, but only for such conifers as *Thuja* and cypress which were relatively easy to strike by the older methods. It did not help at all with the really difficult subjects, such as pines, larches and Douglas fir. With the moderately difficult Sitka spruce it improved take considerably, and encouraged further investigations. It was observed that there were very large differences between individual trees in respect of rooting ability, and these were unrelated to ease of grafting. One hopeful indication was that whereas the original cutting from a mature tree might be difficult, subsequent cuttings from the plant raised in this way appeared easier to root, indicating some degree of reversion to juvenility. In fact, the key to cutting reproduction appears to be the encouragement of shoots or other tissues of juvenile characteristics. There is some hope in the pines of rooting fascicular shoots encouraged by pruning or disbudding, particularly on very young plants, and it is also

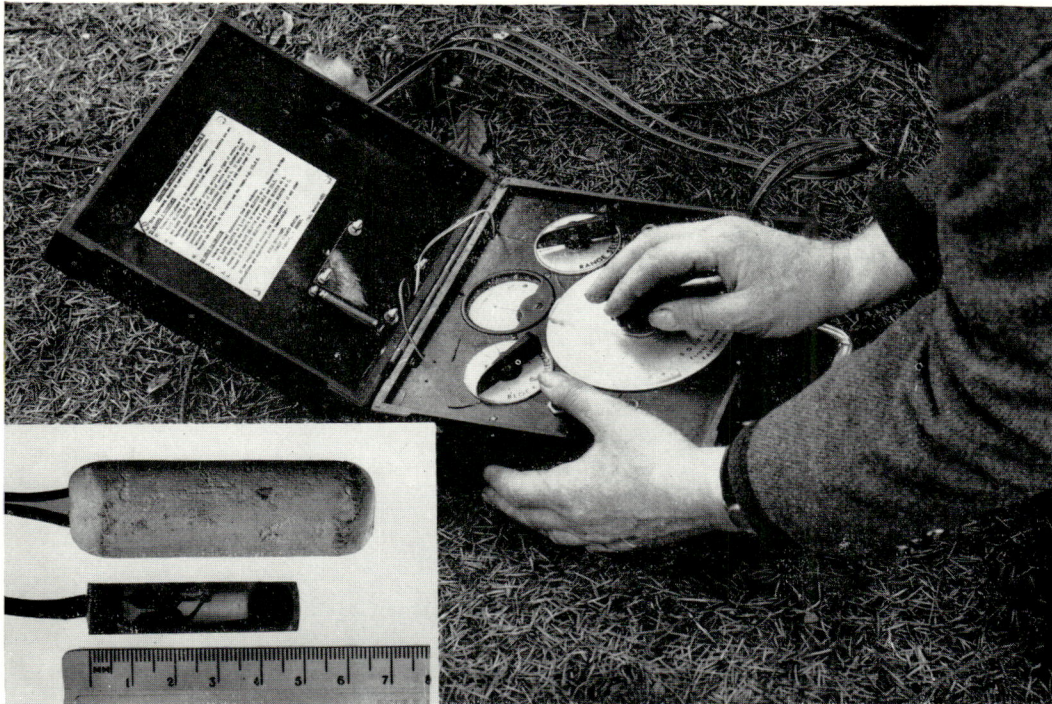

Plate 56. Measuring moisture content and temperature of soil, using a portable alternating current bridge. Inset: Gypsum soil moisture block and encapsulated thermistor. Ch. 23. D5899.

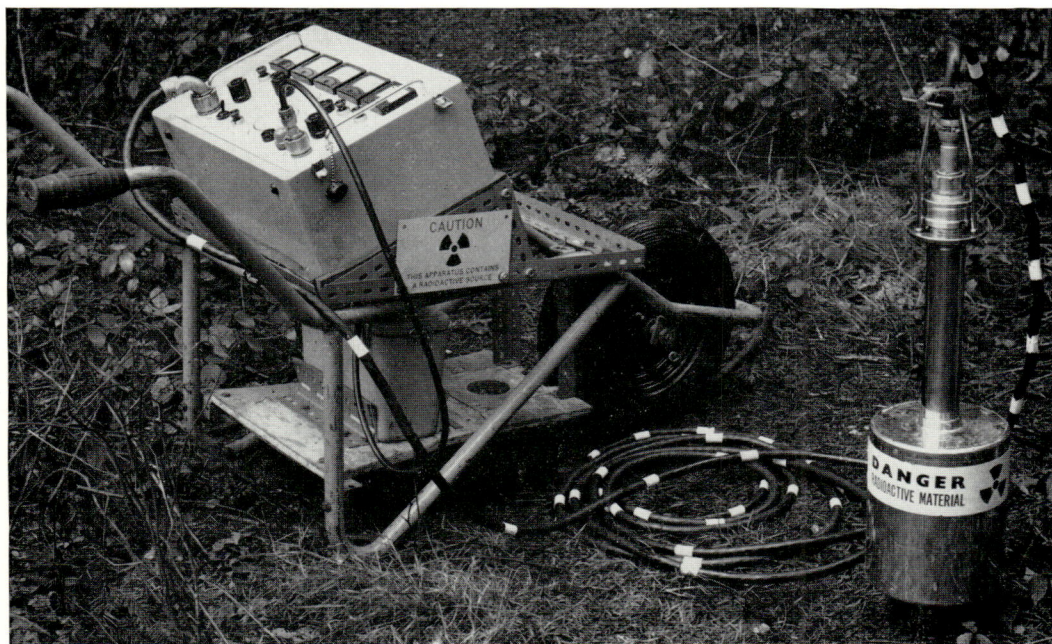

Plate 57. Measurement of moisture content of soil using a neutron probe. The probe (on right of photograph) contains a source of fast neutrons. A detector for slow neutrons is lowered to a pre-determined depth down an access tube. Neutrons emitted by the source collide with the nuclei of atoms in the surrounding soil, their energies are moderated, particularly by hydrogen, their courses are altered and some find their way back to the detector and are recorded. Most of the hydrogen occurs as a constituent of water, and therefore it is possible to relate changes in count rate to change in moisture content. Ch. 23. C3926.

Plate 59. Grafting Plus tree scion on to sturdy rootstock; waxing has been completed. The geneticists working on suitable techniques in vegetative propagation found grafting to be more successful than cuttings. Ch. 22. D5107.

Plate 58. An elite Scots pine in the Black Wood of Rannoch in Rannoch Forest, Perthshire. The natural type of Scots pine particularly desired by the geneticists. Good timber type with straight main stem, definite leader and branches springing at right angles. Good natural cleaning. Very little of this type now left due to the selective felling of the best trees. Ch. 22. B2940.

Plate 61. Stump treatment with fungicide to control *Fomes annosus*. Dr J. Rishbeth showed that the principal means of dispersal in plantations was through the aerial carriage of basidiospores to freshly cut stumps; from these infected stumps the local spread to living trees was by means of root contact. Trials conducted with fungicides led to the more or less general adoption of cut stump protection, using at first creosote, in the late 1950s. Ch. 25. B5684.

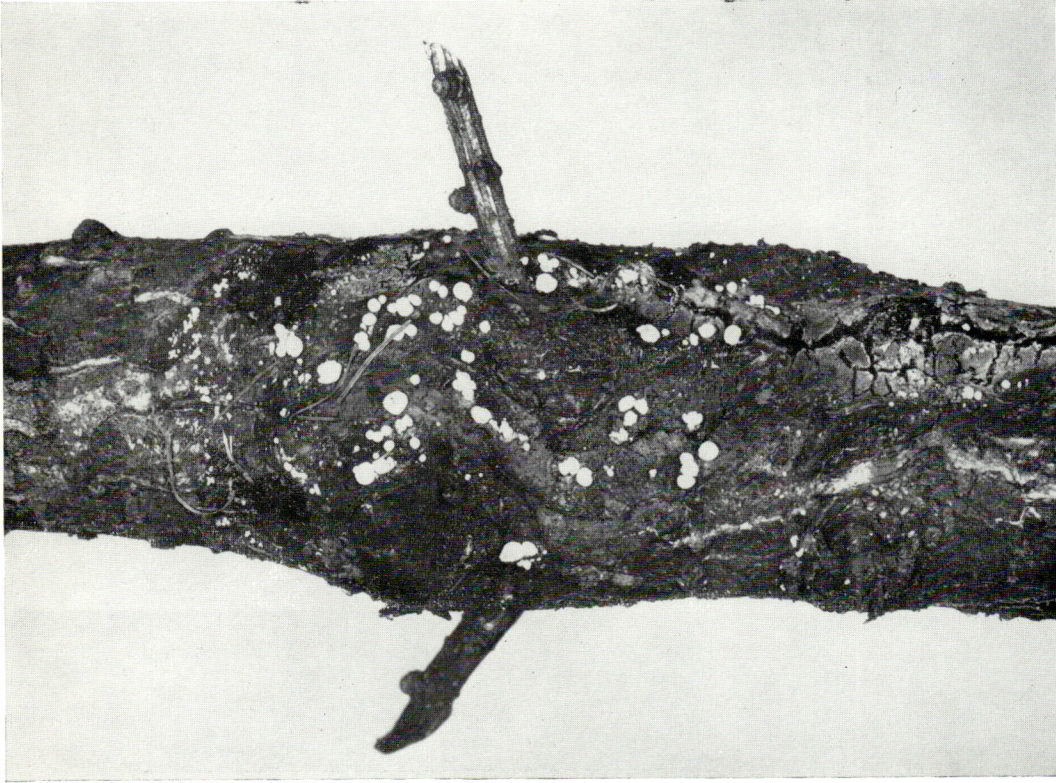

Plate 60. Larch canker, *Dasyscypha (Trichoscyphella) willkommii*, studied by Day and Peace before and after 1940. The association of the disease with provenance was clearly established. Day regarded the fungus as secondary, the primary damage being due to frost. Ch. 7 and 25. D4641.

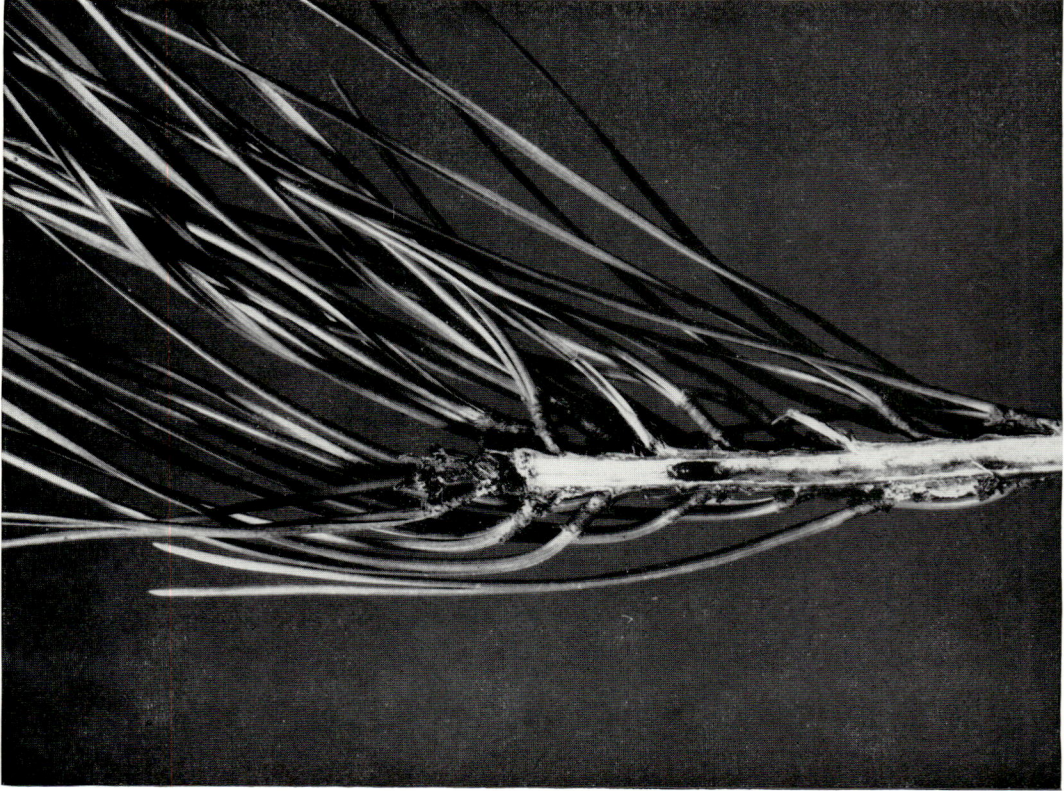

Plate 63. Pine shoot beetle, *Tomicus (Myelophilus) piniperda*, damage to Scots pine at Wareham Forest, Dorset, in 1946, showing the typical spire tops associated with this type of damage. This damage resulted from the large fire in 1946. Ch. 25. D1884.

Plate 62. Pine shoot beetle, *Tomicus (Myelophilus) piniperda*, damage to Scots pine shoot. There were temporary outbreaks in the early 1950s where the conversion of poles from thinnings had been centralised at depots, but this was controlled by rationalising the conversion process to avoid large dumps of unbarked material during the breeding period. *Tomicus* was a greater hazard after the gale of January 1953, but experiments in the use of insecticides on stacked pine timber were successfully developed, using first DDT and later BHC preparations. Ch. 25. C3206.

Plate 65. Insect suction trap developed at Alice Holt Lodge, Hampshire, for collecting segregated periodic samples of flying insects directly into a liquid. Used in the study of aphids and adelgids, this equipment has given a great deal of information on species present and their flight periods. Ch. 25. A3959.

Plate 64. Defoliation of Scots pine by Pine looper moth larvae, *Bupalus piniarius*, at Cannock Chase Forest, Staffordshire. Ch. 25. D3846.

Plate 66. Feeding damage by larvae of the Pine looper, *Bupulas piniarius*, on needles of young Scots pine shoot. Note three characteristic types of feeding: (1) Wasteful feeding of needles by small nibbles up and down the needle. (2) Eating of whole needles down to the base. (3) Eating of needles down to the mid-rib or vein. Populations of this insect, large enough to defoliate pines, first occurred in Cannock Chase Forest, Staffordshire, and the dune plantations of Culbin, Laigh of Moray Forest, Morayshire, in 1953. Aerial spraying took place in the summer of 1954 at both forests, using one pound of DDT per acre. It was the first exercise of this kind undertaken in Britain, and was entirely successful as a short-term measure. Ch. 25. D3999.

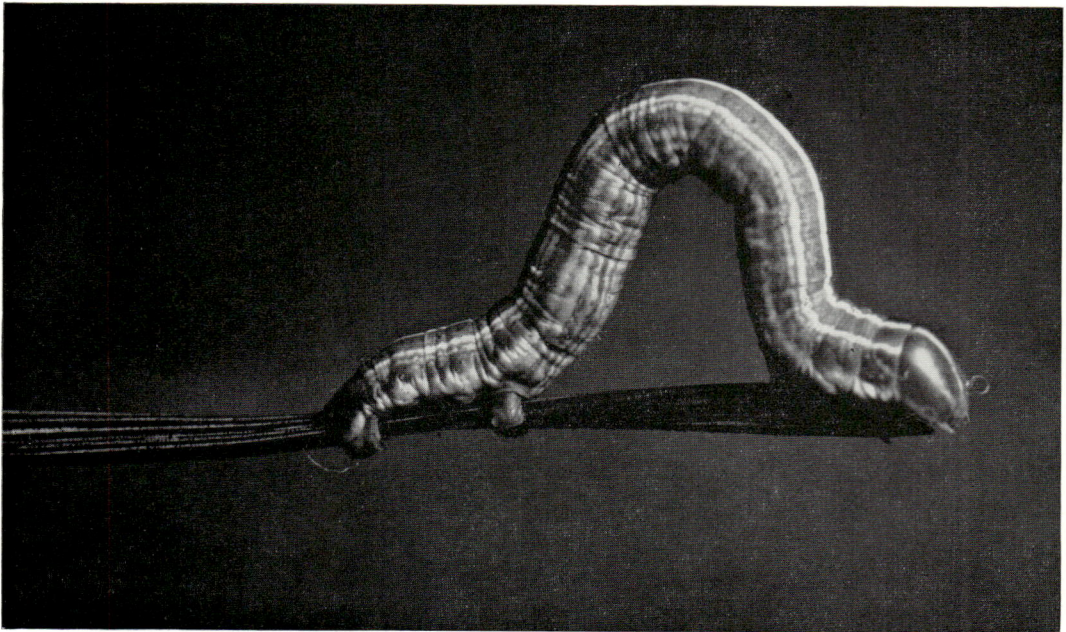

Plate 67. Full-grown larva of the Pine looper moth, *Bupalus piniarius*, on a needle of Scots pine, showing the larva in a typical looper position by which it moves along, bringing its rear claspers up to its thoracic legs. Ch. 25. A1667.

Plate 68. Sitka spruce thinning experiments at Forest of Ae, Dumfriesshire, showing plantations planted at 3 ft × 3 ft (1 × 1 m approx.). Thinning practice in the post-war period attracted much argument, but replicated thinning experiments were rare till the early 1960s. Ch. 26. C4237.

Plate 69. Sitka spruce thinning experiments at Forest of Ae, Dumfriesshire, showing plantations planted at 8 ft × 8 ft (2.4 × 2.4 m approx.). Ch. 26. C4234.

Plate 70. Bitterlich's Relascope in use in the forest. This was introduced in about 1950 and gives an estimate of basal area from a count of the number of stems, in a 360° sweep, that subtend angles over certain values at the instrument. Ch. 26. D62.

Plate 71. The Barr and Stroud Optical Dendrometer developed for accurate diameter measurements of standing trees, probably the most important improvement in instrumentation after the war. Ch. 26. BS.

probable that physiological studies under controlled environmental conditions will pay dividends.

However, there was a sufficient mastery of propagation techniques to commence breeding work as soon as the material could be identified. By 1952 the practice of storing clones in "tree banks" had been commenced, and by 1960 over 1,200 clones were represented in these collections of genetic material. The first seed orchards for Scots pine and for the production of hybrid larch had also been laid down in the fifties; the first seed from a hybrid larch seed orchard being collected in the season of 1956/7. The development of the seed orchard required knowledge of the genotype, assessed in the first place by tests of open pollinated progeny of plus trees and further by deliberate crossings to test combining ability and the heritability of the character selected. Again, the mechanics of the process required a good deal of work—protection of flowers, timing of pollination, storage and testing of pollen, etc. Also the whole question of flowering; its onset and possible control began to assume importance. At this stage basic physiological studies were called for, and in 1956 P. F. Wareing and K. A. Longman (1960) of Manchester University commenced work on the physiology of flowering under Commission grant; the start of a long and profitable association. Their studies threw much light on such questions as the physiological age and size of plants in relation to flowering; photoperiodic effects; and tropisms such as the arrangement of branches and flower inception. Practical investigations on flowering and cone production included nutritional studies and devices known to fruit growers such as girdling, root pruning etc. Girdling in particular was often found to be efficient in inducing coning in trees of mature growth. Pollination is shown on the front cover.

Over the years, and as the numbers of plus trees selected and represented in seed orchards mounted, the task of progeny testing grew to very exacting dimensions; and as in other plant breeding work, highly sophisticated patterns of crossings were adopted. In some cases (notably in beech) the variation between the progeny of individual open pollinated trees was larger than that between widely separated provenances. This of course suggests that the second stage of improvement—mother tree selection—may be a very well worth while one. As Faulkner has remarked, it is easily effected in large-scale plantations of Sitka spruce by felling selected dominants for seed in good seed years; any slight local production or genetic loss (e.g. to natural regeneration) being easily outweighed by even a small percentage gain in productivity in the large area of plantations raised from such seed.

Sitka spruce presents some exacting problems for the breeder. An early objective was to improve resistance to *Elatobium* (*Neomyzaphis*) *abietinum*; it was a common observation that individuals sometimes appeared to escape attack. However, it was found that this phenomenon did not have a simple genetic basis. The most desirable improvement in Sitka spruce is to increase the density of the timber. In the selection of plus trees it is very useful to have some quick means of assessing density; in recent years a technique of density determination from rings of a standard age using increment borings has been worked out at the Forest Products Research Laboratory (Brazier, 1970). The initial selection on form and vigour is in the order of 1 in 750,000 of seed originally sown, and amongst the plus trees identified on external characters only some 15 per cent reach the national average of density. It is obvious that unless the heritability of such a character turns out to be high it will be a slow process to build it in by conventional breeding methods.

Much progress has been made in speeding up progeny trials by growing seedlings under glass with a degree of environmental control (the techniques have recently been described by Herbert, 1971), and if juvenile and adult characters correlate usefully, the long generation period in forest trees can of course be shortened. Obviously where the heritability of desirable characters is low, the temptation must be to go for clonal material; as in poplar breeding, where the process is simple because of the ease of vegetative reproduction. (Plate 33.)

Work in forest genetics in this country has not proceeded long enough to attempt any analysis of prospects for the later stages of tree improvement. It is quite obvious that the gains from the early stages are potentially large, and we may not have cashed in on these to the fullest extent yet.

Reverting to the more general question of seed production; the surveys of genetic material had by 1960 identified over 400 stands extending to 7,000 acres considered suitable for seed collection. At this period there was a strong movement towards the improvement of forest tree seed through control of collection and certification of origin. At the British level this was marked by the formation of two Tree Seed Associations concerned with the certification of origin of seed and plants moving through home nurseries. The basis for certification was, necessarily, the register of seed sources drawn up by the Genetics Section, of which a provisional version was prepared by 1962. The Register was progressively revised during the sixties, and the system of classification of seed sources improved. In its final shape it was designed to comply with the requirements of the international scheme drawn up by the Organisation for Economic Co-operation and Development.

The old system of seed identity numbers (year of collection/arbitrary numeral), which had been in use from 1921, had been replaced in 1957 by a system of permanent numbers based on the Universal Decimal Classification geographic code. The classification of home seed sources was based, primarily, on the level of improvement from the wild; seed from orchards in production occupying the top category, and seed from normal stands the lowest; but practical considerations such as ease of collection were also noted. It must be admitted that the work of identifying and classifying stands got well ahead of progress in organising home seed collection, in spite of useful development work (notably by D. T. Seal) on methods of collection. However by the end of the period it could be said that collection had become a much better organised activity.

REFERENCES

ALDHOUS, J. R. (1962). Provenance of Sitka spruce: An account of the nursery stage of experiments sown in 1958. *Rep. Forest Res., Lond.* 1961, 147–154.

BRAZIER, J. D. (1970). Timber improvement. II. The effect of vigour on young-growth Sitka spruce. *Forestry* **43** (2), 135–150.

EDWARDS, M. V. (1955). Norway spruce provenance experiments. *Rep. Forest Res., Lond.* 1954, 114–126.

EDWARDS, M. V. (1956). The design, layout and control of provenance experiments. *Z. Forstgenet. ForstpflZücht.* **5** (5/6), 169–180.

FAULKNER, R. (1962). Seed stands in Britain and their management. *Q. Jl For.* **56**, 1–15.

FAULKNER, R. (1965). Seed orchards in Britain. *Rep. Forest Res., Lond.* 1964, 211–243.

HERBERT, R. B. (1971). *Development of glasshouse techniques for early progeny test procedures in forest tree breeding.* Forest Rec., Lond. 74.

LIGHTLY, A. A., and FAULKNER, R. (1963). Grafting conifers at Grizedale nursery. *Q. Jl For.* **57**, 293–301.

LINES, R. (1966). Choosing the right provenance of Lodgepole pine. *Scott. For.* **20**, 90–103.

LINES, R., and MITCHELL, A. F. (1968). Provenance: Sitka spruce. *Rep. Forest Res., Lond.* 67–70.

MATTHEWS, J. D. (1962). *Seed collection and tree breeding in Britain.* Pap. 8th Br. Commonw. For. Conf. East Africa.

WAREING, P. F., and LONGMAN, K. A. (1960). Studies on the physiology of flowering in forest trees. *Rep. Forest Res., Lond.* 1959, 109—110.

Chapter 23

ECOLOGICAL, BOTANICAL, AND PHYSIOLOGICAL STUDIES

The grouping here is not particularly logical, but serves to bring together certain work in the botanical sciences, some of which is basic, and none strictly applied. It has been mentioned in Chapter 11 that a small section in Forest Ecology under J. M. B. Brown was maintained over the post-war period; the appointment to Commission Research of a physiologist was on the other hand a late development. However, P. F. Wareing's grant-aided studies in tree physiology were started in the mid-fifties, and some support was also given to the programme in forest hydrology at Oxford under L. Leyton: there was also collaboration with A. J. Rutter of Imperial College in his work on this subject. Besides assisting Rutter, the Soil Section at Alice Holt had its own work on moisture and tree growth. There has been no appointment of a Forest Botanist as such, and taxonomic work at a practical level has been undertaken by a number of people in the Research Branch; some special studies have been made with Commission support by botanists in the Universities.

Forest Ecology
J. M. B. Brown's work had two main aspects; firstly autecological studies of particular trees, and secondly environmental studies, local or more general. In the first category, his most important work was on beech, based on a detailed study of beechwoods throughout the country (Brown, 1953). This gave a very clear picture of the characteristics of beech stands, whether spontaneous or planted, in relation to soil and other environmental factors, and also provided much information on the silviculture of the tree, with special regard to the regeneration of beech woodland and its establishment in various environments. He did not neglect the genetic factor in the quality of beech stands. He followed beech with studies on Corsican pine, or perhaps more correctly *Pinus nigra*, of which Corsican pine is the most important geographical variety.

This proved a more difficult subject to round off, since this exotic had been planted very widely, and several pathological symptoms were associated with unsuitable environments: there was also the provenance story. At a fairly early stage Brown's surveys went a long way to defining the safe, productive limit of this tree. Corsican pine has high summer heat requirements, and in the cooler parts of the country and at high elevations becomes increasingly sensitive to elevation, unfavourable aspects, and ill-drained soils. It is open to question how much study should be devoted to exotics where it is reasonably clear that they have been carried outside their useful range;

enough must of course be done to suggest the nature of their limits.

Local site studies have usually been undertaken in connection with some difficulty in afforestation, or failure of a species in a particular environment. Brown's preliminary investigations usually gave an idea of the limiting factors, and gave pointers for experimental treatments. Brown made more general studies on the chalk, on coastal environments (especially dunes), and latterly became engaged in the difficult task of disentangling the factors behind the relatively poor growth of Sitka spruce in parts of South Wales, where it has often been suspected that industrial pollution is accentuating the effects of exposure and soil deficiencies. Much of Brown's work depended on detailed observation, especially of soil and rooting characteristics, but he also made instrumented studies on light, temperature (especially the incidence of frosts in relation to cover) and evaporation.

Little ecological work of a general nature was supported by the Commission at outside institutions in the post-war period. Exceptions were a study of the site factors influencing the regeneration of ash by P. Wardle (1957) of the School of Botany, Cambridge, and work on the light requirements of the important conifers by W. A. Fairbairn of Edinburgh University from 1964 (Fairbairn and Neustein, 1970).

Botanical Studies
The principal object in morphological work (especially in the conifers) has been to find characters distinguishing geographical variations and provenances. E. V. Laing (1950–61) at Aberdeen Universiy did a great deal of work in this field in the forties and fifties with Commission support. Besides investigating racial and provenance variations in a number of theimportant conifers (larch, Douglas fir, Scots pine, Lodgepole pine, etc.) he studied the characters of the hybrid larch *L.* x *eurolepis* and other conifer hybrids. A. Carlisle followed his work on Scots pine and made a comprehensive study of the variations of indigenous pine (which provided base material for *The native pinewoods of Scotland*, in which he collaborated with H. M. Steven) (Steven and Carlisle, 1959).

The distribution of the two native oaks, pedunculate and sessile, has always been a matter of interest. In the sixties J. E. Cousens and D. C. Malcolm (1962–65) studied the distribution and morphological variations of these two oaks. They found very great variation especially in pedunculate oak, so much as to cast doubt on its status as a good botanical species. Sessile oak appeared to be better defined, but pure populations were rare, and it seemed very possible that introgression with pedunculate oak had occurred.

Commission workers followed lines similar to Laing's; for instance, Edwards and Lines paid a good deal of attention to the variation in Lodgepole pine, and Buszewicz studied seed and young seedling characters (cotyledon number, etc.) to find criteria for checking declared seed origin in the laboratory. Much of this work dealt with continuous variation, and any useful distinctions found were expressible as population statistics, rather than the "either-or" characters favoured by the botanist. In fact foresters' observations of variation have very often emphasised its continuity, and have frequently cast doubt on the accepted definition of species or variety.

Physiology

This has been one of the most active and rewarding divisions of the plant sciences in the post-war period, and in forest research there has been much interest in the extension of physiological studies to trees. Physiological research is basic by definition, since it aims at the understanding of the fundamental growth processes. It may however be of great assistance in solving short-term problems, and much of the work mentioned here has had practical objectives.

It should be mentioned that tree physiology was one of the topics chosen by the Forestry and Woodlands Research Committee of the Natural Environment Research Council for special attention, and the support given by the Commission up to 1965 was later much expanded by the Council.

The first of the grant-aided studies in tree physiology were commenced by Prof. P. F. Wareing of Manchester University (and later Aberystwyth) in 1956. This arose directly from one of the most important practical problems in J. D. Matthew's breeding programme; the length of the generation in forest trees. K. A. Longman (who incidentally held the post of Physiologist in the Research Branch from 1970 to 1972) and subsequently L. W. Robinson collaborated with Wareing in studies of juvenility and the onset of maturity up till 1963 (Wareing and Longman, 1957–60; Wareing and Robinson, 1961–1963). Two broad approaches were made, firstly to find what governs the onset of maturity (flowering and seed bearing) in forest trees, and secondly to induce flowering in trees which were physiologically ready for it.

The former topic is of course a very large and complex story, but some very useful indications were obtained at an early stage. Perhaps the most important of these was to dissociate maturity from any simple concept of age, either in calendar years, or induced periods of growth; in birch for instance

maturity seemed to be governed more by size, since seedlings grown continuously under long day conditions began to flower when nine feet tall, but only thirteen months old. Subsequent work on a number of other woody species supported the idea that a minimum size had to be reached for maturity and this encouraged efforts to secure rapid early growth in seed orchards.

Grafting experiments with various species using mature on juvenile stocks and vice versa showed that these states had some degree of persistence; they were certainly not immediately reversed according to the state of the stock. Observations on the flowering morphology of larches suggested that geotropic effects had something to do with the onset of flowering, and branch training experiments confirmed this; it is apparently the basis for certain fruit pruning practices.

In 1961 N. G. Smith collaborated with Wareing on basic studies in the rooting of cuttings, using the easily rooted poplar *Populus* 'Robusta' as an experimental subject (Wareing and Smith, 1963–65). These studies yielded a good deal of useful general information. Firstly they showed that in the leafless cutting, the active (non-dormant) bud was of importance. Pre-chilling (or of course normal winter conditions) relaxed dormancy. In summer wood cuttings the leaves (growing or no) influence rooting; and it was shown that leaves or active buds were responsible for a build up of auxins at the cutting base.

Later in the period Wareing and D. R. Causton started work on assimilation and dry matter production with broadleaved species (birch and sycamore) as experimental subjects. These studies, albeit preliminary, were of much interest in showing the relationship between leaf area and net assimilation rate in dry matter production, and especially the compensatory mechanism in birch by which individual leaf area and net assimilation rate increases markedly in debranched plants (Causton and Wareing, 1967). Work on these lines may be expected to follow a number of paths. For example there is the detailed analysis of stands as productive systems, and work of this nature is now being undertaken under the auspices of the Natural Environment Research Council. Possibly of greater importance is the study of the growth habits in individual trees, which may assist in the recognition of highly productive genotypes.

Hydrological Studies

Work on the water economy of forest trees has two aspects: it is an integral part of the study of the tree's physiological requirements, and there is also the major hydrological aspect, where the water balance of stands or catchments is the principal interest.

Forest research is interested in both, but in this country the primary concern in hydrological work has lain with the Hydrological Research Unit, which was established under the Department of Scientific and Industrial Research in 1962 and is now one of the divisions of NERC. Forest hydrology has a long history, but in Britain had not attracted much attention before 1956. Then F. Law's paper to the British Association at Sheffield suggested that the water consumption of spruce greatly exceeded that of grasslands, which was not the general opinion at this time (Law, 1956). Law's contribution raised a lively controversy over the merits or otherwise of forest cover in catchment areas, and was undoubtedly an important stimulus towards hydrological research.

The Hydrological Research Unit has concentrated on factors governing the gross water balance of catchments; climate, soils, land usage etc; and has established several catchment experiments of the type developed in America and South Africa. The Commission, whilst not concerning itself in hydrological work as such (except for assistance to the Unit in the provision of sites etc.), was associated with work on the hydrological elements carried out by L. Leyton and E. R. C. Reynolds (1961–71) at Oxford, and by A. J. Rutter (1967) of Imperial College. Rutter's work was not grant aided, but the Soil Section collaborated with him in his field work, which was mainly carried out at Crowthorne in Bramshill Forest, Berkshire.

Leyton and Reynolds investigated the measurement of rainfall over and in forest stands, including stem flow. They also studied ways of estimating evapo-transpiration, concentrating on direct methods based on the moisture movement in stems (such as the heat flow method); in this they used the intriguing device of enclosing the whole tree in a plastic envelope and measuring the moisture exchange in the artificially circulated atmosphere as a means of calibrating the stem flow determinations. They also made special studies of interception of rainfall in various types of ground vegetation, and investigated moisture distribution in the soils under forest stands.

Rutter was chiefly interested in evapo-transpiration. His main method was to determine by moisture gauges the dates at which soils left and regained field capacity; the measured precipitation between these dates being taken as the transpiration plus the evaporation of intercepted rainfall. On deep sandy soils suited to moisture gauging, and in a dry locality with normally long periods between the appearance of deficits in spring and rewetting in late autumn or mid-winter, there was good reason to suppose that this method gave reliable estimates. Rutter's values for evapo-transpiration of Scots pine crops exceeded those for neighbouring grassland

considerably, and were slightly higher than the potential evapo-transpiration for a free water surface calculated by Penman's formula (E_0). Rutter and the Oxford workers emphasised the role of interception in the evaporation of forest stands: it was shown that intercepted rain was in fact evaporated from foliage at a much greater rate than occurs in transpiration, and hence there is not full compensation. This supported Law, but it has been felt that the rather small plantation in which he made his determinations exaggerated the evaporation rates by an "oasis" effect.

W. H. Hinson and R. L. Kitching started work on moisture and tree growth at Alice Holt in the early sixties (Fourt and Hinson, 1970). Study plots similar to Rutter's were laid down at Bramshill, Hampshire, in Corsican pine and Douglas fir, and the changing patterns in soil moisture observed throughout the season by a series of moisture blocks inserted down the profile to a depth of eight feet. The soil drying patterns under the two species differed considerably, Corsican pine drying out the soil to a much greater depth than Douglas fir, and consequently a good deal more rainfall was required to rewet the soil under the pine. Much the same results were obtained in a comparison between these two species at Aldewood, Suffolk, and it must be assumed that Douglas fir is more economical in its water use under these relatively dry conditions; an intriguing result, since, if thinking generally about the behaviour of the two species in Britain, most people would guess that Douglas fir had the greater moisture requirement. A similar study comparing Sitka spruce and Corsican pine in the New Forest failed to detect any great difference in transpiration between the spruce and the pine.

Other southern soils were investigated. Very marked differences from these sands were observed in the moisture patterns on the Chalk and the Chalky Boulder Till. Here there were no clear gradients in moisture tensions with depth, and it was evident that movements of water upwards in the chalk were very significant.

One of the aims of this work had been to study the relationship between diameter growth and soil moisture and for this purpose vernier bands were used to detect small girth changes. This enquiry gave surprisingly negative results. Some small departures from the march of girth increment were almost certainly due to changes in stem hydration, but there was relatively little relationship between girth increment during or between seasons and soil moisture tension, even when plots were subjected to artificial drought by "roofing" the forest floor.

Though not a primary object of work on soil moisture, these enquiries gave a useful opportunity to make a comparison with the evapo-transpiration estimates obtained by Rutter at Crowthorne, when in the seasons of 1964/66 a period of 19 months elapsed between leaving field capacity and full rewetting in the Corsican pine plots at Bramshill. The rainfall for this period (assumed equal to the total evapo-transpiration) was extremely close to the Penman estimate for potential evaporation from a free water surface (E_o) and allowing for seasonal differences, a rather lower value than Rutter's, but not by an important margin.

The instrumentation of the soils moisture work received a good deal of attention. Kitching and Hinson improved the circuitry for measuring the electrical resistance in gypsum moisture blocks, and a very convenient and useful tubular tool was developed for taking soil samples for gravimetric work. The neutron moisture meter was introduced in 1965, and this made it much easier to carry out intensive sampling of soil moisture patterns. Kitching improved the method of determining the moisture status of tree stems by the measurement of electrical resistance between electrodes inserted in the tree; this detected sizeable differences in moisture content which were more or less in line with soil moisture tensions. (Plates 56 and 57.)

Some of the work on moisture approached climatology. From 1961 till the end of the period a type of Garnier gauge designed by F. H. W. Green of the Nature Conservancy was maintained at Alice Holt. These simple, watered, grassland lysimeters gave values for evaporation (rainfall plus added water less drainage) which were usually in close accord with potential evapo-transpiration (E_t) calculated by Penman's formula, especially for the whole season, and incidentally all experiences appear to confirm the usefulness of this formula.

By the end of the period the work of the Soils Section on moisture was shifting to the wetter environments, where problems are concerned with surpluses rather than deficits. The Section had already made important contributions towards drainage in theory and in instrumentation, and one might expect intensification of this side of the work.

REFERENCES

BROWN, J. M. B. (1953). *Studies on British beech-woods*. Bull. For. Commn, Lond. 20.

CAUSTON, D. R., and WAREING, P. F. (1967). Influence of leaf-characters and growth habit on the production of dry matter by forest trees. *Rep. Forest Res., Lond.* 1966, 103; 1967, 161.

COUSENS, J. E., and MALCOLM, D. C. (1962–1965). Oak population studies. *Rep. Forest Res., Lond.* 1961, 109–113; 1962, 119–120; 1963, 112–117; 1964, 127–129.

FAIRBAIRN, W. A., and NEUSTEIN, S. A. (1970). Study of response of certain coniferous species to light intensity. *Forestry* 43 (1), 57–71.

FOURT, D. F., and HINSON, W. H. (1970). Water relations of tree crops: A comparison between Corsican pine and Douglas fir in South-east England. *J. appl. Ecol.* 7, 295–309.

HINSON, W. H., and KITCHING, R. (1964). A readily constructed transistorised instrument for electrical resistance measurements in biological research. *J. appl. Ecol.* 1, 301–305.

KITCHING, R. (1965). A precision portable electrical resistance bridge incorporating a centre zero null detector. *J. agric. Engng Res.* 10, 264–266.

KITCHING, R. (1966). Investigating moisture stress in trees by an electrical resistance method. *Forest Sci.* 12 (2), 193–197.

LAING, E. V. (1950–1961). Studies on the morphological variations of conifers. *Rep. Forest Res., Lond.* 1949, 78; 1950, 124; 1951, 135; 1952, 120–121; 1953, 121–122; 1954, 56–57; 1955, 79–80; 1956, 95–96; 1957, 108–109; 1960, 99–101.

LAW, F. (1956). The effect of afforestation upon the yield of water catchment areas. *J. Br. WatWks Ass.* 38, 489–494.

LEYTON, L., and REYNOLDS, E. R. C. (1961–1971). Hydrological relation of forest stands. *Rep. Forest Res., Lond.* 1960, 101; 1961, 100–101; 1962, 112–113; 1963, 107–108; 1964, 116–120. With F. B. THOMPSON: *Rep. Forest Res., Lond.* 1965, 119–123; 1967, 160; 1969, 158–161; 1970, 187–189; 1971, 142–144.

RUTTER, A. J. (1967). Forests and evapotranspiration: An analysis of evaporation from a stand of Scots pine. In *Forest Hydrology* (ed. W. E. Sopper and H. W. Lull), pp. 403–417. Oxford: Pergamon Press.

STEVEN, H. M., and CARLISLE, A. (1959). *The native pinewoods of Scotland*. Edinburgh: Oliver & Boyd.

WARDLE, P. A. (1957). Notes on the ecology of ash. *Rep. Forest Res., Lond.* 107–108.

WAREING, P. F., and LONGMAN, K. A. (1957–1960). Studies on the physiology of flowering in forest trees. *Rep. Forest Res., Lond.* 1957, 106–107; 1958, 107–109; 1959, 109–110.

WAREING, P. F., and ROBINSON, L. W. (1961–1963). The juvenility problem in woody plants. *Rep. Forest Res., Lond.* 1960, 95–96; 1961, 114–115; 1962, 125–127.

WAREING, P. F., and SMITH, N. G. (1963–1965). Physiological studies on the rooting of cuttings. *Rep. Forest Res., Lond.* 1962, 122–125; 1963, 118–121; 1964, 130–132.

Chapter 24

FOREST SOILS

It has been mentioned that the Commission singled out forest soils research for special attention in the post-war period. Such a wide field can of course be approached in many ways, but in the wartime deliberations most interest was expressed in the evolution of forest soils on heathlands and moorlands; especially on those where for one reason or another afforestation had been a difficult process. There was a good deal of uncertainty about the direction of the changes to be expected following canopy formation; whether for instance the podsolisation typical of *Calluna*-dominated vegetation would be arrested under conifers, or whether changes might even be for the worse. Nor were the underlying biological processes at all well known for these British conditions. At the practical level, all might be boiled down to the long-term maintenance or enhancement of fertility, but while this was open to silvicultural experimentation, it was rightly considered that the whole subject needed underpinning by basic research.

The Imperial (now Commonwealth) Forestry Institute, with interests wider than purely home forestry, strengthened its research capacity in the immediate post-war period, and set up a small team to specialise in various aspects of forest soils. In furthering its policy to encourage soils work, the Commission provided the funds for two scientists attached to the Institute for several years, working in close contact with Prof. Champion's team. At the same time the old Macaulay Institute grant arrangement was broadened to allow full time attention to Scottish soils problems. A Commission-supported appointment was also made at Rothamsted.

Of these workers, P. J. Rennie at Oxford and J. D. Ovington at the Macaulay Institute concentrated on physical and chemical problems of heathland and newly afforested soils (the early Macaulay work being centred on the fixed dunes at Culbin, Laigh of Moray Forest, Moray), whilst P. W. Murphy at Oxford and A. C. Evans at Rothamsted started investigations into the roles of the soil micro-arthropods in litter breakdown and humus formation (Murphy, 1950–59; Evans and Guild, 1947–48). In addition to this direct investment in basic soils research, support was given from time to time to the permanent staff of the Institute (usually in the form of small grants for assistant staff, and practical help in the field), and hence the Commission kept closely in touch with the work of G. W. Dimbleby, L. Leyton, and W. R. C. Handley; covering the ecological aspects of soil evolution; soil physics and chemistry; and the biochemistry of litter breakdown

respectively. Most of the Oxford work was carried out in connection with the soils at Allerston and other important silvicultural experimental areas in Yorkshire.

The establishment of the Nature Conservancy in 1949 brought in a new and independent interest in woodland soils and J. D. Ovington's work on the effects of different forest trees on the site, which stemmed from his earlier work at the Macaulay Institute, was largely carried out in Commission forests (Ovington, 1953–58). The Forestry Commission opened its own Soils Laboratory at Alice Holt in 1955, enabling analytical and instrumented studies of soils to be undertaken in addition to the qualitative observations on profiles which were an important part of the Ecologist's site studies. In 1953, the Commission appointed a small sub-committee of the Research Advisory Committee to assist in coordinating work on forest soils; this was initially chaired by Sir William Gammie Ogg, the other members being G. V. Jacks, Prof. H. G. Champion, and Prof. H. M. Steven. From the early fifties most support of basic soils work was through normal grant procedure rather than "outstationing". During the latter half of the post-war period, Edinburgh University took up soils micro-biological studies under D. Gifford and A. J. Hayes, and for some years in the late fifties and early sixties work at Bangor on the subject of mycostasis was also supported by the Commission.

Only a somewhat cursory review of this diverse field of work will be attempted, and it should be emphasised that direct and grant-supported Commission work does not constitute the whole story by any means in this particular field.

Basic soils research in the period has concentrated on soil biology and the nutritional cycle, with a lesser amount of work on moisture which has been more conveniently included with hydrology in Chapter 23.

Soil Biology and Biochemistry

The long-term objective in these studies has been to get an understanding of the agencies and processes involved in the evolution of forest soils, especially those processes leading to the development of mull and mor types of humus, as distinguished in the classical studies of P. E. Muller in the late nineteenth century.

Murphy, Evans and later Gifford concentrated on soil micro-fauna, the mites (*Acarina*) being the creatures most closely studied (Gifford, 1959–63; 1964–65). Early work was perforce exploratory, a great deal of patient effort going to finding ways of extracting them from the litter, and to identifying and enumerating the populations. Later it became possible to study feeding preferences and the relative distribution of the various species down the profile, and to compare populations in different environments. Gifford for instance studied the micro-fauna of the grass *Molinia coerulea* litter, as compared with that of young spruce plantations, and of older Scots pine woodlands. Both Murphy and Gifford were interested in the relations between fungal agencies and mite feeding in litter breakdown, and in recent years Hayes (1964–67) has made a special study of fungal colonisers of Scots pine litter. Another invertebrate group of some importance in litter breakdown, the *Collembola*, have also received attention from Murphy, and from Poole (1957) at Bangor.

Most of the above work was concerned with mor-type litter, but Heath (1961–66) at Rothamsted worked on broadleaved litter (oak and beech) in the south, especially on the role of earthworms which are primary feeders on leaves in mull-type soils. He obtained much information on their rates of feeding and preferences, oak leaves being usually preferred to beech and "shade" leaves of either species to the harder "sun" leaves. By protecting leaves in nylon mesh he was able to estimate the part played by earthworms in the breakdown of litter compared with alternative feeders such as *Collembola* and enchytraid worms.

Handley's preliminary studies on mull and mor formation pointed to the constituents of the living plant as the primary cause of these distinct forms of humus; he considered the micro- and meso-fauna as secondary agencies. Over a certain range of soils mul or mor appeared to be reversible conditions. He concentrated on *Calluna* as the dominant ground vegetation on podsolised soils, and his analytical work indicated that a substance, allied to the tannins, present in the leaves of *Calluna*, protected the proteins from decomposition, thus leading to the build-up of humus rich in nitrogen, but in unavailable forms (Handley, 1954). The further study of these tannins was the subject of grant-aided research in collaboration with Prof. B. R. Brown and colleagues at the Dyson-Perrins Laboratory, Oxford, in the early sixties. This work examined the tannins in the leaves of a number of species in relation to the actions of known enzymes concerned in the breakdown of plant substances. (Brown and Love, 1960–1962; Bocks et al., 1963–64).

Handley also studied the mechanism of the well-known *Calluna* effect on spruces and other sensitive species, and put forward the hypothesis that the basic cause was the inhibition of mycorrhizal association in the tree species due to some exudate from the endophyte in *Calluna*. At the chemical level, Leyton had produced good evidence that the effect was to starve the sensitive tree species of nitrogen.

Dobbs and his colleagues' work at Bangor on mycostasis, which was supported by the Commission in the late fifties, belongs on the edge of this field (Dobbs et al., 1959–63). From these studies, it appears that the property of certain soils by which the germination of fungal spores is inhibited is of bacterial origin.

G. W. Dimbleby's main field of work in palaeoecology, which has done much to clarify the history of certain important soils, falls outside the scope of this publication; but he conducted several field experiments in conjunction with other members of the Oxford team and Commission Research staff at the Allerston group of forests, near Scarborough in Yorkshire. His chief interest in this was to establish an ecological equilibrium at a higher level, using broadleaved species to reverse the current trend towards increasingly severe podsolisation.

The Nutritional Cycle

While the biological studies mentioned above had little or no immediate effect on applied research, there has been no hard and fast line between basic work on the cycle and applied work on improved nutrition. The main agencies for Commission-supported work in the period were the Macaulay Institute and the Imperial Forestry Institute, but the Rothamsted/Forestry Commission programme on nursery nutrition was also concerned with the nutrition of young plants in the forest. The Soils Section at Alice Holt (the laboratory opened in 1955) undertook both practical experimental and objective-basic work in this field, besides studies on moisture which are mentioned in Chapter 23.

P. J. Rennie (1950–64; 1955) concentrated on the physical and chemical aspects of the heathland soils at Allerston, being specially interested in the capability of the soils to support the growing crop. Rennie's work on aeration of heathland soils strongly supported deep cultivation. Determinations of the main elements in the rootable profile compared with estimates of crop uptake suggested that these poor soils might fail to support rapid growth of conifers in later life. Rennie thought calcium might prove limiting before phosphorus or potash, but there has been little experimental evidence to suggest this. However, Rennie's investigations were undoubtedly influential in directing thought to the nutrition of the growing crop, as distinct from fertilisation purely for the establishment phase. His work did not suggest that broadleaved soil improvers by themselves could do much good on basically poor soils.

J. D. Ovington at the Macaulay Institute started

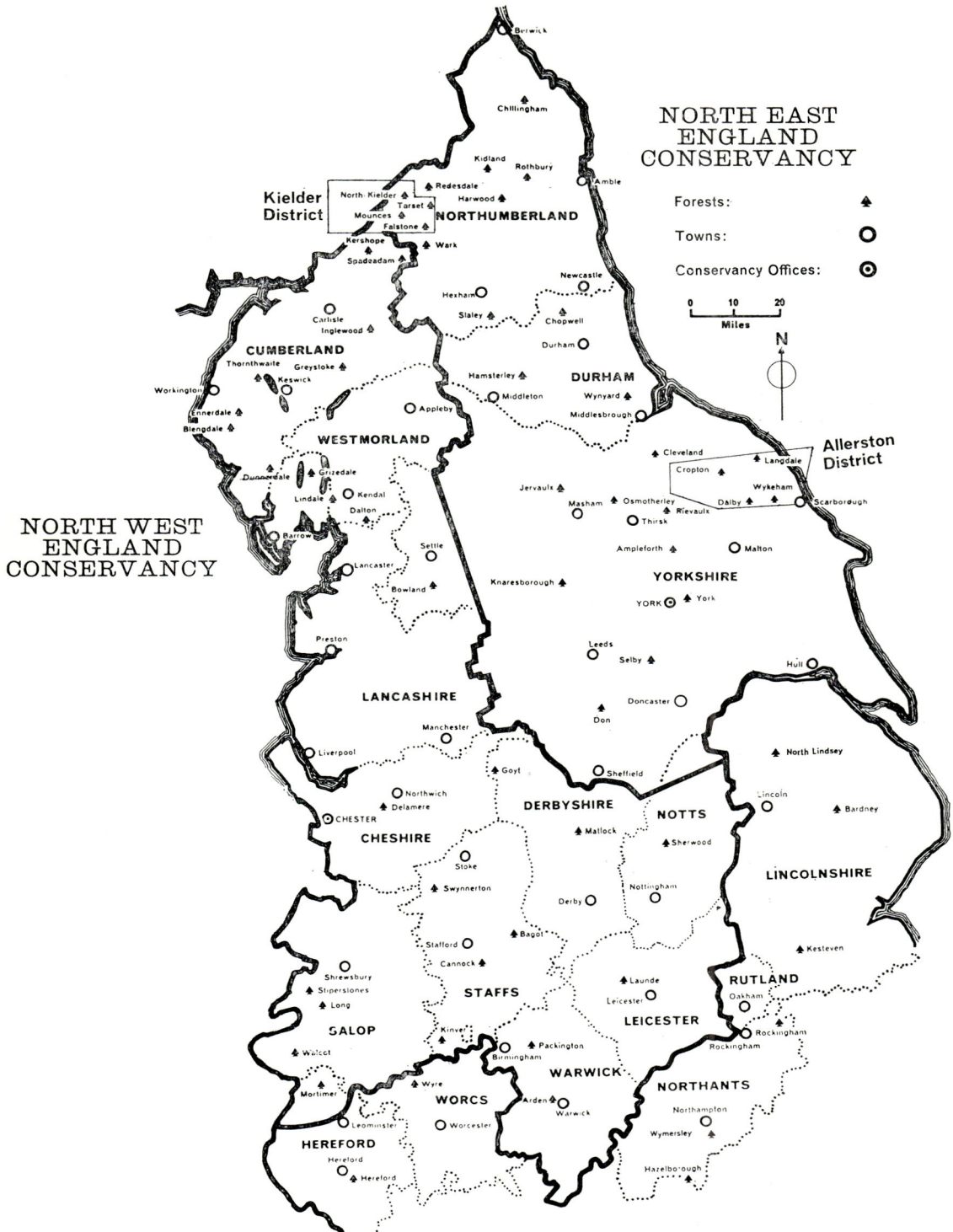

NORTH EAST
ENGLAND
CONSERVANCY

Kielder
District

NORTHUMBERLAND

Forests: ♠

Towns: ○

Conservancy Offices: ◉

0   10   20
Miles

N

Berwick

Chillingham

Kidland   Rothbury

North-Kielder   Redesdale
Tarset   Harwood   Amble
Mounces
Falstone
Kershope   Wark
Spadeadam

Carlisle
Inglewood

Newcastle

Hexham

Slaley   Chopwell

CUMBERLAND

Thornthwaite   Greystoke
Keswick

Workington

Ennerdale
Blengdale

Durham

Hamsterley   DURHAM

Middleton   Wynyard
Middlesbrough

Appleby

Cleveland   Allerston
District
Cropton   Langdale
Wykeham
Jervaulx   Dalby   Scarborough
Masham   Osmotherley   Rievaulx
Thirsk

WESTMORLAND

Dunnerdale   Grizedale
Lindale   Kendal
Dalton
Barrow

Settle

Ampleforth   Malton

YORKSHIRE

Knaresborough

YORK   York

Lancaster

Bowland

Leeds
Selby

Preston

LANCASHIRE

Doncaster
Don

Manchester

Liverpool

North Lindsey

Goyt   Sheffield

Northwich
Delamere

CHESTER

DERBYSHIRE   NOTTS

Lincoln
Bardney

CHESHIRE

Matlock

LINCOLNSHIRE

Stoke

Swynnerton

Sherwood

Derby   Nottingham

Kesteven

Stafford
Cannock

Bagot

Shrewsbury
Stipersiones
Long

STAFFS

Launde
Leicester

RUTLAND
Oakham

Kinver

LEICESTER

Rockingham
Rockingham

SALOP

Walcot

Packington
Birmingham

WARWICK

NORTHANTS

Mortimer   Wyre

WORCS

Arden
Warwick

Northampton

Leominster

Worcester

Wymersley

HEREFORD

Hereford

Hereford

Hazelborough

NORTH WEST
ENGLAND
CONSERVANCY

Plate 72. Map: Commission Forests in Northern England.

Plate 73. Map: Commission Forests in Southern England and Wales.

EAST ENGLAND CONSERVANCY

DERBYSHIRE

- Goyt
- Matlock
- Sherwood
- Derby ○
- Nottingham ○
- Bagot

NOTTS

LINCOLNSHIRE

- North Lindsey
- Lincoln
- Bardney
- Kesteven

TAFFS

RUTLAND

- Laude
- Leicester
- Oakham
- Rockingham
- King's Lynn

LEICESTER

HUNTS

- Packington
- Birmingham
- Huntingdon

WARWICK

NORTHANTS

- Arden
- Warwick
- Northampton
- Wymersley

CAMBS

○ **CAMBRIDGE**

- Bedford
- Hazelborough
- Buckingham
- Ampthill

BEDS

OXON

BUCKS

HERTS

- Bernwood
- Oxford
- Kennington Nursery
- Chiltern
- Rothamsted
- Bramfield
- Hertford
- Walden

ESSEX

- Chelmsford

BERKS

- Hungerford
- Devizes
- Reading
- Andover
- Bramshill

□ **LONDON**

- Rochester

Cromer ○

- Wensum

NORFOLK

- Norwich ○

- Thetford Chase
- Thetford
- Southwold ○

- Bury St. Edmunds

SUFFOLK

- Aldewood
- Ipswich
- Lavenham

- Colchester

- Shipbourne

KENT

- Canterbury

HANTS

- Micheldever
- Basingstoke
- Guildford
- Dorking
- Redhill

SURREY

- Abinger
- Tunbridge Wells
- Ashford
- Challock
- Dover

- Winchester
- ALICE HOLT
- Hursley
- Petersfield
- Horsham
- St. Leonards

LYNDHURST

- Southampton
- Queen Elizabeth Forest
- Arundel

SUSSEX

- Lewes
- Bedgebury
- Hastings

- New Forest
- Chichester
- Brighton
- Eastbourne

- Wight
- Newport

N

SOUTH EAST ENGLAND CONSERVANCY

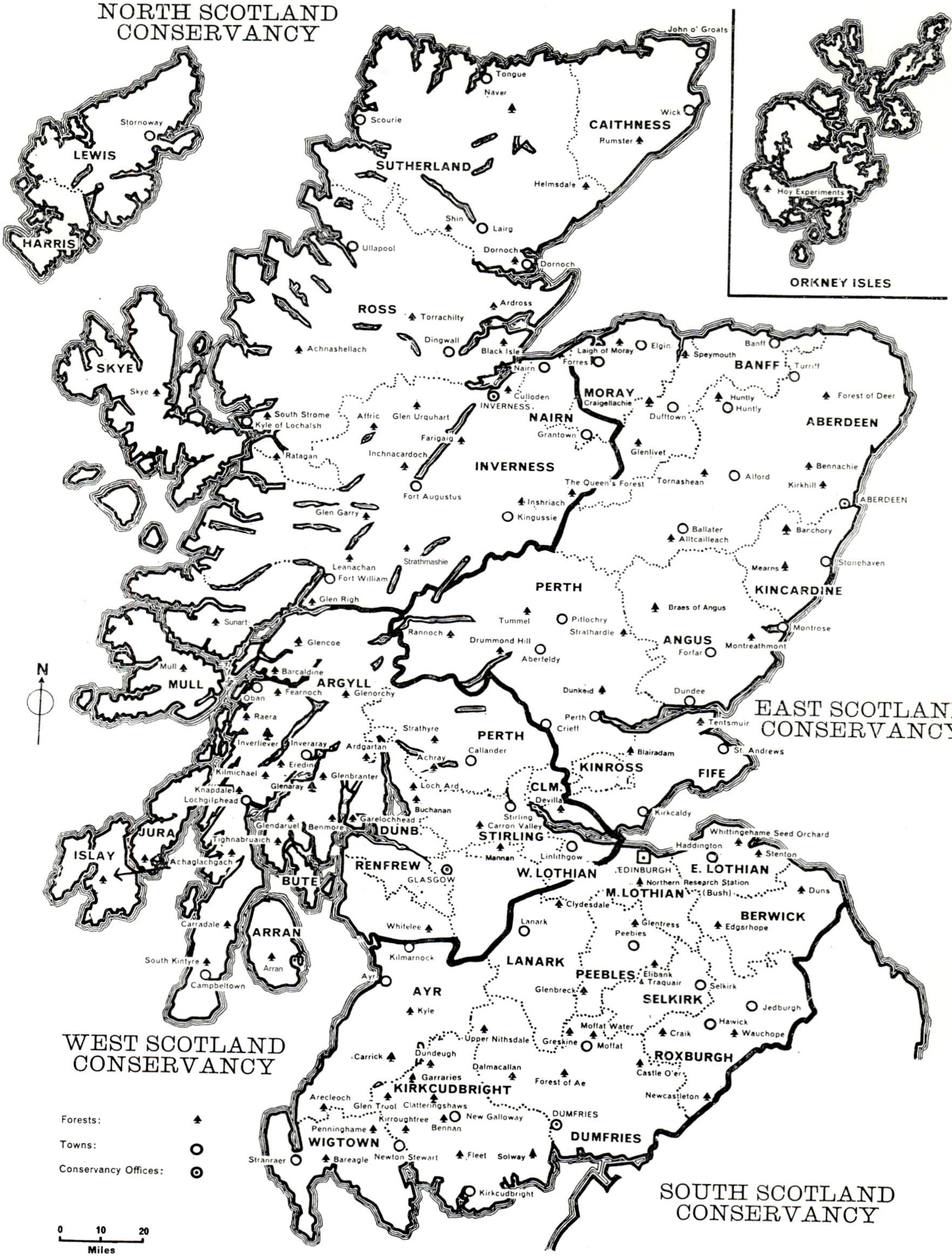

ORKNEY ISLES

John o' Groats

LEWIS

Stornoway

HARRIS

CAITHNESS

Tongue
Naver
Wick
Scourie
Rumster
SUTHERLAND
Helmsdale

Hoy Experiments

Shin
Lairg
Ullapool
Dornoch
Dornoch

ROSS
Ardross
Torrachilty
Achnashellach
Dingwall
Black Isle
Laigh of Moray
Elgin
Banff
SKYE
Nairn
Forres
Speymouth
Turriff
Culloden
BANFF
Skye
INVERNESS
MORAY
Craigellachie
Huntly
Forest of Deer
South Strome
Affric
Glen Urquhart
NAIRN
Dufftown
Huntly
Kyle of Lochalsh
Farigaig
Grantown
ABERDEEN
Ratagan
Inchnacardoch
INVERNESS
Glenlivet
Alford
Bennachie
The Queen's Forest
Tornashean
Kirkhill
Fort Augustus
ABERDEEN
Glen Garry
Inshriach
Ballater
Kingussie
Alltcailleach
Barchory
Strathmashie
Mearns
Stonehaven
Leanachan
KINCARDINE
Fort William
PERTH
Braes of Angus
Montrose
Glen Righ
Tummel
Pitlochry
Sunart
Rannoch
Strathardie
ANGUS
Glencoe
Drummond Hill
Forfar
Montreathmont
Barcaldine
Aberfeldy
MULL
Fearnoch
ARGYLL
Dunkeld
Dundee
Glenorchy
Oban
EAST SCOTLAND
Raera
Perth
CONSERVANCY
Strathyre
Crieff
Inverliever
Inveraray
PERTH
Tentsmuir
Ardgartan
Callander
St. Andrews
Eredine
Achray
Blairadam
Kilmichael
Glenbranter
KINROSS
FIFE
Knapdale
Glenaray
Loch Ard
Lochgilphead
Buchanan
CLM.
Kirkcaldy
JURA
Glendaruel
Benmore
Garelochhead
Devilla
Whittingehame Seed Orchard
Tighnabruaich
DUNB.
Stirling
Haddington
ISLAY
Achaglachgach
Carron Valley
Stenton
BUTE
STIRLING
W. LOTHIAN
E. LOTHIAN
RENFREW
Mannan
Linlithgow
EDINBURGH
Carradale
GLASGOW
M. LOTHIAN
Northern Research Station
ARRAN
W. LOTHIAN
(Bush)
Duns
South Kintyre
Arran
Clydesdale
BERWICK
Campbeltown
Whitelee
Lanark
Glentress
Edgarhope
Kilmarnock
Peebles
LANARK
PEEBLES
Elibank
AYR
Ayr
Traquair
Selkirk
Kyle
Glenbreck
SELKIRK
Jedburgh
Moffat Water
Hawick
Greskine
Moffat
Craik
Wauchope
Carrick
Dundeugh
Upper Nithsdale
ROXBURGH
Dalmacallan
Castle O'er
Garraries
Forest of Ae
KIRKCUDBRIGHT
Newcastleton
Arecleoch
Glen Trool
Clatteringshaws
Forests:
Penninghame
Bennan
New Galloway
DUMFRIES
Towns:
Kirroughtree
DUMFRIES
Conservancy Offices:
WIGTOWN
Bareagle
Newton Stewart
Fleet
Solway
Stranraer
Kirkcudbright

0   10   20
Miles

Plate 74. Map: Commission Forests in Scotland.

the investigations which were developed by T. W. Wright and subsequently by W. O. Binns, H. G. Miller and others. These centred firstly on the fixed dunes at Culbin, Moray, where pine plantations were entering the canopy stage. In this low rainfall area and on these permeable sands moisture might be expected to be a limiting factor, and moisture stress determinations down the profile showed large seasonal deficiencies under pine crops, whereas unplanted dunes showed little drying (Ovington, 1950). Wright was able to show later that deficiencies were smaller under heavily thinned than lightly thinned crops (Wright, 1955–56). Later developments along these lines (mentioned in Chapter 23) succeeded in quantifying the water economy of pine crops, but it may be remarked here that it never proved easy to demonstrate a limiting effect of moisture on relatively deep-rooted pine crops under British conditions.

The main theme of the Macaulay work at Culbin was the cycling and distribution of nutrients in the soils and the crop. Ovington's first studies showed that there were large changes in the distribution of the main nutrient elements in the profile following afforestation; noticeable ten years after planting and very markedly after twenty years. Nitrogen, phosphorus, potash and calcium were depleted in the lower profile and much concentrated in and just below the litter layer. After joining the Nature Conservancy he made comparisons of the soil conditions under a wide range of species in the Bedgebury Forest Plots and in species trials on other types of soil. He also estimated the nutrients in the standing crops. Whilst some overall distinction could be made between broadleaved trees and conifers, the variants inside these groups were very great, and some conifers such as Douglas fir, the larches, *Thuja plicata* and Lawson cypress had foliage comparatively rich in minerals, and had not the strong mor-forming litter of the pines.

Wright obtained estimates of the major nutrient elements in all parts of the pine stand. An important point (on a poor soil such as the Culbin sand) was the relatively large proportion of the elements in the tree/soil system contained in the wood and bark, and thus normally removed in harvesting (Wright and Will, 1958). He also studied seasonal variation in foliar contents, giving useful guidance for timing sampling in diagnosing deficiencies (Wright, 1957).

A logical development was to study the uptake and fate of added nutrients in conjunction with the measurement of growth responses and this approach was also followed by W. H. Hinson at Alice Holt on southern manurial experiments. A notable feature was the large quantity of nitrogen (in unavailable state) building up in the litter and humus, but H. G. Miller also showed that substantial quantities of added nitrogen were stored in the tree for three or more seasons: the tree itself might be the more effective reserve for this element (Miller, 1969).

In addition to the studies at Culbin, the Macaulay carried out basic investigations on the peats, the old experimental centre of the Lon Mor being a useful site. W. O. Binns studied the striking effects of the first successful Lodgepole pine, *Pinus contorta*, plots on deep peat. Marked shrinkage with deep fissuring had occurred following the drying out of the peats, and some part of the change was thought to be irreversible. Binns looked at the depletion of nutrient by the crop, and found evidence that potash might be critical on such sites (Binns, 1968). From the early sixties onwards the Macaulay work was increasingly concerned with the nutrient requirements of forest stands. By this time the methods and interpretation of foliar analyses had been brought to a useful level, and analyses were being increasingly applied in diagnosis of deficiencies.

The Soil Laboratory at Alice Holt adopted Macaulay methods of soil and foliar analyses, but Hinson was responsible for several improvements in technique. Besides providing a growing analytical service, mainly to Silviculture, several investigations of a basic nature were carried out. Work on moisture was mentioned in the preceding chapter. Much attention was given to methods of sampling both in the crop and in the tree. Some interesting observations were made on the fate of added phosphorus to the extremely deficient spruce crops at Wilsey Down, Kernow Forest, Cornwall, much the greatest proportion being held superficially in the inverted turves, though the spruce itself had retained a good deal more than the ground vegetation. A general enquiry into the relationship between quality class of spruce and the fertility of the site, in which whole-tree analysis was being used to gauge the nutritional level of the crop, was unfortunately not completed due to the pressure of service work. The Soils Laboratory graduated into an independent Section when W. O. Binns joined the Commission on leaving the Macaulay, and from then took over the experimental work on forest nutrition in the south.

Such work has always included studies of the efficiency of different compounds or formulations in supplying particular elements; uptake and plant response being the usual criteria. There has however been interest in the effects of different fertilisers, especially nitrogenous compounds, on the litter itself; storage, rate of humification etc. It has been claimed that free ammonia or compounds releasing nitrogen in that form were less subject to leaching in forest soils than the more commonly used salts, and Hinson found some evidence to support this (Hinson and Reynolds, 1958). In 1961 J. Tinsley of the Department for Soil Science at Aberdeen laid down experiments in Scots pine crops to study the physical and chemical effects of different forms of nitrogen on the soils and litter; liming being also included. Tinsley's studies indicated that the differences between ammonia and ammonium sulphate were short-lived, it was not easy to distinguish effects from forms (or even rates) after four or so seasons; the effects of lime on the other hand were quite discernible in higher carbohydrate content in the humus (Tinsley, 1963–68).

As would be expected, research on the chemical aspects of the nutritional cycle has had more practical application than the biological studies, since the latter deal with processes which are secondary and governed by the energy and nature of the food source, and are not by themselves easily manipulated. There has probably been a change in thought about forest litter during the period; there is not now so much emphasis on the evils of mor, and this is due in part to the recognition of *Calluna* humus as a special case. However the mor characteristics of uneven distribution of nutrients in the profile, and especially the large stores of unavailable nitrogen, pose many problems for the future. Most of these revolve round the central question of soil conservation, and are essentially long-term. There will be a great stimulus to further soil biological investigations should it appear that we cannot keep the system going by purely cultural methods, and as certain monocultures in the southern hemisphere do seem to have run into troubles, second and third generation crops in this country will be carefully watched. So far, however, there has been little suggestion of the "replant" problem which occurs with *Pinus radiata* and other pines in Australia and Africa.

REFERENCES

BINNS, W. O. (1968). Some effects of tree growth on peat. *Proc. 3rd int. Peat Congr.*, 358–365.

BOCKS, S. M., BROWN, B. R., and HANDLEY, W. R. C. (1963–1964). The action of enzymes on plant polyphenols. *Rep. Forest Res., Lond.* 1962, 93–96; 1963, 88–94.

BROWN, B. R., and LOVE, C. W. (1960–1962). Substances in leaves affecting the decomposition of litter (Protein-fixing constituents of plants). *Rep. Forest Res., Lond.* 1959, 104–109; 1960, 102–106; 1961, 90–92.

DOBBS, C. G., BYWATER, J., and GRIFFITHS, D. A. (1959–1963). Studies in soil mycology. *Rep. Forest Res., Lond.* 1958, 98–104; 1959, 94–100; 1960, 87–92; 1961, 95–100. With N. C. C. CARTER: 1962, 103–112.

EVANS, A. C., and GUILD, W. J. M. (1947–1948). Studies on the relationships between earthworms and soil fertility. *Ann. appl. Biol.* 34, 307–330; 35, 1–13, 181–192 and 471–493.

GIFFORD, D. R. (1959–1963). Soil fauna research. *Rep. Forest Res., Lond.* 1958, 116; 1959, 100–102; 1960, 98–99; 1961, 94–95; 1962, 98–103.

GIFFORD, D. R. (1964–1965). Studies of soil micro-arthropod populations in Scottish forests. *Rep. Forest Res., Lond.* 1963, 164–172; 1964, 101.

HANDLEY, W. R. C. (1954). *Mull and mor formation in relation to forest soils.* Bull. For. Commn, Lond. 23.

HAYES, A. J. (1964–1967). Studies on the mycology of Scots pine litter. *Rep. Forest Res., Lond.* 1963, 96–97; 1964, 102–110; 1965, 107–109; 1967, 147–148.

HEATH, G. W. (1961–1966). Biology of forest soils (Soil faunal investigations). *Rep. Forest Res., Lond.* 1960, 94–95; 1961, 92–94; 1962, 96–97; 1963, 94–96. Earthworms: 1964, 96–100; 1965, 103–106.

HINSON, W. H., and REYNOLDS, E. R. C. (1958). Cation adsorption and forest fertilisation. *Chemy. Ind.* 7, 194–196.

MILLER, H. G. (1969). Nitrogen nutrition of pines on the sands of Culbin Forest, Morayshire. *J. Sci. Fd Agric.* 20, 417–419.

MURPHY, P. W. (1950–1959). Soil faunal investigations. *Rep. Forest Res., Lond.* 1949, 67–71; 1950, 113–116; 1951, 130–134; 1952, 123–126; 1953, 110–116; 1954, 60–61; 1955, 83–84; 1956, 91–94; 1957, 109; 1958, 106–107.

OVINGTON, J. D. (1950). Afforestation of the Culbin sands. *J. Ecol.* 38, 303–319.

OVINGTON, J. D. (1950–1951). Forest soil investigations in Scotland (Ecological studies in pine plantations). *Rep. Forest Res., Lond.* 1949, 75–76; 1950, 121–124.

OVINGTON, J. D. (1953–1958). Studies of the development of woodland conditions under different trees. *J. Ecol.* 41, 13–34; 42, 71–80; 43, 1–21; 44, 171–179 and 597–604; 46, 127–142 and 391–405.

OVINGTON, J. D. (1959a). Mineral contents of plantations of *Pinus sylvestris* L. *Ann. Bot.* 23 (89), 75–88.

OVINGTON, J. D. (1959b). The circulation of minerals in plantations of *Pinus sylvestris* L. *Ann. Bot.* 23 (90), 229–239

OVINGTON, J. D. (1961). Some aspects of energy flow in plantations of *Pinus sylvestris* L. *Ann. Bot.* 25 (97), 12–20.

OVINGTON, J. D. (1962). The nutrient cycle and its modification through silvicultural practice. *Proc. 5th Wld For. Congr.*, Seattle 1960, 533–538.

OVINGTON, J. D., and PEARSALL, W. H. (1956). Production ecology. II. Estimates of average production by trees. *Oikos* 7 (2), 202–203.

POOLE, T. B. (1957). Soil collembola in a Douglas fir plantation. *Rep. Forest Res., Lond.* 109–111.

RENNIE, P. J. (1950–1964). Research into the

physical and chemical properties of forest soils. *Rep. Forest Res., Lond.* 1949, 71–75; 1950, 116–118; 1951, 129–130; 1952, 108–116; 1953, 107–110; 1954, 62–63.

RENNIE, P. J. (1955). The uptake of nutrients by mature forest growth. *Pl. Soil* 7 (1), 49–95.

TINSLEY, J. (1963–1968). Chemical changes in forest litter. With R. J. HANCE: *Rep. Forest Res., Lond.* 1962, 128–130; with A. LENNOX: 1963, 97–107; 1964, 111–115; with A. A. HUTCHEON: 1965, 110–118; 1967, 166.

WRIGHT, T. W. (1952–1960). Influence of tree growth on soil profile development. *Rep. Forest Res., Lond.* 1951, 126–127; 1952, 116–117; 1953, 103–105; 1954, 53–54; 1955, 73; 1956, 81–82; 1957, 96–97; 1958, 105–106; 1959, 102–104.

WRIGHT, T. W. (1955–1956). Profile development in the sand dunes of Culbin Forest, Morayshire. I. Physical properties. *J. Soil Sci.* 6, 270–283; II. Chemical properties. *J. Soil Sci.* 7, 33–42.

WRIGHT, T. W. (1957). Abnormalities in nutrient uptake by Corsican pine growing on sand dunes. *J. Soil Sci.* 8, 150–157.

WRIGHT, T. W., and WILL, G. M. (1958). The nutrient content of Scots and Corsican pines growing on sand dunes. *Forestry* 31, 13–25.

Chapter 25

FOREST PROTECTION

General

As mentioned in Part I, Chapter 1, during the pre-war period the Forestry Commission relied largely on University staff for advice and for investigations in the specialised divisions of forest protection, pathology and entomology. One of the advantages expected in setting up the Research Station at Alice Holt was the closer relationship between such specialists and workers in the more general field of research. It was not considered necessary to set up a general section in Forest Protection, but the Pathologist's coverage has always been much wider than mycology, and has included damage from inorganic agencies.

No arrangements were made for research in forest fires at this time, and in fact this never became a central research activity; most of the advances in fire control methods being due to field officers, notably C. A. Connell, but some work on retardants etc. was done by Silviculture in collaboration with him. However, during the sixties important basic work on forest fuels was carried out at the Joint Fire Research Organisation, Boreham Wood, by arrangement with the Commission.

The higher animals had not figured much in forest research before the war (Elton's work on voles being an exception), but there were several changes in status of animals in the forest in the post-war period which necessitated enquiry. The Grey squirrel rose to an important agricultural and forestry pest, and it became appreciated that we had to live with deer in the forests, and minimise damage by managing the herds. These were not purely research concerns by any means, but the research element was enough to justify a small Sub-Section in Zoology, led by Judith Rowe, which was set up in 1961 under the Entomologist.

The two research Sections of Pathology and Entomology remained small in numbers of scientific staff over most of the pre-war period, though, as in all Commission Research, the adaptability of the Research Foresters was a great strength in field investigations. However, these Sections had always the primary duties of "watchdogs" (as Peace put it) and advisers, and this put limits on the amount of research they could do themselves. More specifically the limitation was in the number of pests or diseases which could be investigated at depth, for at all times their programmes held numerous items where the impact and control of known organisms was under study.

Pathology and Entomology had a common interest in preventing new pests and diseases from entering the country in nursery stocks, seed, or by other means, and they therefore assisted the appropriate Government Departments in drawing up schedules under the Orders governing the import of plant material etc. They were also concerned with quarantine arrangements for special lots of plants imported from abroad. T. R. Peace defined the task of the pathologist as the "avoidance and control of disease", and emphasised the priority of avoidance; in this he summed up the philosophy of both Sections.

Forest Pathology

T. R. Peace held the post of Forest Pathologist from its inception till 1959, when he succeeded M. V. Laurie as Chief Research Officer for a tragically brief period. As already mentioned, he worked with W. R. Day at Oxford before the war, and served as a field officer with the Commission during it.

The greatest achievement of Peace's career at Alice Holt was the publication of his comprehensive work *Pathology of trees and shrubs* (Peace, 1962), which ranks as a major contribution to forest literature.

Peace was followed as Forest Pathologist by J. S. Murray, who left to join the staff of Aberdeen University in 1961. Dr D. H. Phillips succeeded Murray, and served as Forest Pathologist to 1967, being then promoted to Chief Research Officer, South. D. A. Burdekin was then promoted to lead this Section and continued in charge to the end of the period reviewed.

In one of his early reviews Peace commented on the number of diseases which had appeared in crops in the thicket and pole stages. He mentioned the death of Scots pine on chalk, Dieback in larch, Dieback in Corsican pine, Group dying of spruces, "Top dying" of Norway spruce, and (not least) Group dying of pines in East Anglia due to the fungus *Fomes annosus*. This short but diverse case list gives as good a starting point as any other, since it provides an interesting sample of pathological enquiries and their relationship with forestry practice.

Failure of Scots pine on chalk can be dismissed briefly. It had been associated with rendzinas and other soils with much free lime by Day and Sanzen-Baker (1939), and was plainly an extreme form of the nutritional disorder lime-induced chlorosis. The most interesting recent observation on its occurrence in the field is the significance of even very thin acid layers over calcareous soils in preventing the disorder (see also Chapter 16). Canker of European larch associated with the fungus *Trichoscyphella* had been studied by Day (1932; 1950) and Peace (amongst others) before the war, and by the late forties the association of the disease in its most serious form with seed strains of high alpine origin was abundantly clear from the early provenance experiments. Day regarded the fungus *Trichoscyphella* as secondary, the primary damage being due to frost; thicket stage plantations in certain topographical situations being specially prone to frost from the damming of cold air. The role of *Trichoscyphella* in canker formation and dieback has remained somewhat controversial, but J. G. Manners (1953; 1957), who worked in the forties with Commission support at Southampton University, was able to show that the fungus certainly had a degree of pathogenicity, and that infection and cankers could be produced in the absence of frost. Larch canker is a good example of the practicability of avoiding disease—in this case by attention to provenance. Dieback in Corsican pine has shown some features similar to larch canker, for there is an association with low temperature (autumn rather than spring) and a fungus (*Brunchorstia destruens*) of questionable pathogenicity. D. J. Read (1966; 1967) of Hull University showed that the trouble was most prevalent on northern aspects, and also that *Brunchorstia* was a fungus active at low temperatures. Again it is a disease which can be avoided, by restricting the range of the species, and possibly also by choice of variety. (Plate 60.)

Group dying of Sitka spruce, a trouble associated with plantations in the thinning stage, provided a most interesting investigation. The better known pathogens were dismissed, and Day's suggestion that the disease was primarily a matter of failure of "supply" did not seem plausible for many of the sites. One feature noted by J. S. Murray (Murray and Young, 1961) was that old fire sites were very commonly associated with groups of dead trees. These had usually been very small fires for burning slash, or brewing tea. The fungus *Rhizina undulata* had previously been observed in cases of group dying, though it was not thought to have much pathogenic significance. It was however established that this fungus, which has a liking for the sites of brushwood fires in woodlands, can act as a pathogen, and was the main cause of the trouble. The remedy was obvious enough.

Top dying of Norway spruce has proved a much more complicated story. From the absence of any important pathogens, it has always appeared to be a physiological trouble, but root and soil conditions appeared to rule out drought. It was noted, however, that the trouble was often associated with sudden changes in the environment, such as heavy thinning or (more commonly) with exposure of a plantation margin, which might be supposed to have increased transpiration stress.

In the sixties it was found that there was a connection between the onset of the disease (usually apparent as retarded growth) and mild winters, which suggested that the predisposing factor might be high respiration during periods of low assimilation, leading to depletion of carbohydrate reserves. This theory is not established at the time of writing, but there is some evidence in favour. Though the disease in not perhaps of major importance, its solution might be a pointer to one rather general cause of the poor performance of Continental conifers, such as many pines, in the most maritime British climates.

The group dying of pines in East Anglia due to *Fomes annosus* began to assume serious proportions in the mid-forties after the thinning had started in the Thetford plantations dating from the early and mid-twenties. *Fomes* was known from European experience to be a hazard of old arable sites, especially on calcareous soils, a fair proportion of Thetford falling into this category. The killing of young trees is not the most important role of the fungus, which overall causes greater losses as a stem rot of conifers other than pines. It was however fortunate that the disease should turn up in its more dramatic guise so close to Cambridge, for Dr. J. Rishbeth's work (1950–63) on *Fomes annosus*, with Commission support, proved one of our most paying investments in "objective basic" research. In brief, he showed that the principal means of dispersal in plantations was through the aerial carriage of basidiospores to freshly cut stumps, and from these infected stumps local spread to living trees was by means of root contacts. This pointed to the protection of the cut stumps as a means of preventing the entry of *Fomes* to uninfected or lightly infected plantations. He also found that *Fomes* was often replaced by saprophytic fungi where the conditions were suitable; the scarcity of the soil-living fungus *Trichoderma viride* in calcareous soils at Thetford was thought to facilitate the spread of *Fomes* between root systems. Experiments on stump protection laid down by Rishbeth between 1947 and 1949 included the use of creosote, and also biological control by the fungus *Peniophora gigantea* which was suspected to be antagonistic to *Fomes*. The promise of creosote lead to large-scale trials conducted from Alice Holt, and to the more or less general adoption of cut stump protection using creosote in the late 1950s.

After an interval abroad, Rishbeth reopened his work on *Fomes* in 1955, being joined later by D. S. Meredith (1959; 1960), D. Punter (1962–64), G. W. Wallis (1960; 1961), and J. N. Gibbs (1967; 1968). Search was made for a fungicide, less persistent and more selective than creosote, since it was desirable that saprophytes should enter cut stumps as soon as possible. Several substances showed promise, sodium nitrite appearing the best, and this to a large extent replaced the creosote in the late sixties. It was however considered objectionable on some special sites (catchments etc.); search has continued for a perfectly safe substitute and urea has largely replaced sodium nitrate. (Plate 61.)

Further work was done on *Peniophora* which began to show very great advantages (though restricted to pines); once established in the stump, damage from extraction etc. did not remove the protection, and it was not so critical a matter to treat stumps immediately after felling, since *Peniophora* was able to catch up on recent *Fomes* infections. Early work was done with fresh cultures, but by 1963 Rishbeth had prepared tablets containing oidia, and these gave better results, besides being more convenient. By 1968 *Peniophora* in such form was available commercially, and had become the standard treatment in the extensive East Anglian pine forests. Naturally the success of *Peniophora* on pines has encouraged the search for other saprophytes capable of playing the same role on other conifers, and this is proceeding. But this example of biological control of a fungus in forestry must be rare, if not unique.

Besides collaborating with Rishbeth, D. H. Phillips (1963) and his colleagues worked on several other aspects of *Fomes*. Countrywide surveys were carried out on the incidence of *Fomes* in connection with stump protection programmes. It was noted that the spread of *Fomes* was limited on the peats, and also on old broadleaved soils, but here the freedom was only temporary, as conversion to conifers and changes in the soil microflora provided suitable conditions for the build up of the fungus. An important experimental line was the treatment of infected sites for second-rotation crops. Measures applied in the mid-fifties included poisoning and girdling of standing trees (to reduce reserves) prior to felling, and the removal of stumps from clear felled areas. On very badly infected sites in the Thetford area the last treatment was the only one to give a useful measure of control.

Surveys also gave useful information on the economic significance of rot and on the relative susceptibility of various conifers; though more reliable information will be obtained later from trials of species laid down specifically for this purpose. In the meanwhile there is evidence that *Abies grandis* is not very susceptible (supporting the reputation of *Abies* species in this respect) but *Tsuga heterophylla* and *Thuja plicata* appear highly susceptible, and the larches and Norway and Sitka spruces also became badly decayed.

Less has been done on the other root and stem rotting fungi, of which *Armillaria mellea* and *Polyporus schweinitzii* are the commonest. The former belongs in old broadleaved woodland, and does not appear significant in coniferous forest, but *Polyporus schweinitzii* is considered of some importance as a stem rot of conifers, and may require detailed study. Stem rot caused by the fungus *Stereum* was shown to be associated with damage to butts and major roots in extraction of produce.

While stem rot of living trees is clearly in the forest pathologist's province, the deterioration of *felled* timber in the forest brings in the forest products specialist and the entomologist. Blue stain (Holtman, 1968) in pines is one of the commonest causes of degrade, and is very much a borderline topic, since it brings in such questions as date of felling in relation to insect attack as well as conditions for the spread and development of the fungus causing the stain. A recent joint study between the Research Branch and the Forest Products Research Laboratory has made it clear that the main danger period in felling is May to July; the traditional remedy is of course winter felling and fast extraction, but it has been shown that when logs are felled in the danger period a combined fungicidal and insecticidal treatment gives a very useful measure of control.

Besides larch canker (mentioned above in connection with dieback), half-a-dozen or so other stem or bark diseases have received attention. Much the best known of these is the Dutch elm disease, on which Peace worked before the war, during the first serious epidemic.

During most of the post-war period, the disease was fairly quiescent, for no very obvious reason, but there were one or two active periods following (it is thought) good breeding seasons for the beetle vectors. Peace discontinued his surveys of affected areas in 1949, and his account of the disease was published as Forestry Commission Bulletin No. 33, *The status and development of elm disease in Britain* (Peace, 1960). Any effective countrywide control of the disease, either through the fungus or its vector, offers peculiar difficulties, and the only practical measure appeared to be the removal of dead and dying material. Obviously the best line is selection and breeding for resistance, and as mentioned in Chapter 15, the Research Branch collaborated with the Dutch workers in the selection of breeding material amongst English elms, and in testing Dutch and home selections for resistance. At the time of writing, 1972, the disease has reached another peak of activity, and is causing much alarm.

Bacterial canker of poplars attracted a considerable amount of work over the period, and as with elms, varietal resistance was the main line; the programme being shared between Silviculture and Pathology. From the pathological point of view, the main advances lay in methods of testing for resistance. Planting clonal trial material close to cankered sets did not answer well, and was quickly superseded by inoculation with natural bacterial slime, collected in spring when the flow is most copious. Ratings for resistance and recommendations for planting during the post-war period were based on such tests. The front-cover picture shows inoculation.

Following Ridé's success in isolating *Aplanobacterium*, Continental practice changed to inoculation with pure culture, and Burdekin (1972) adopted this method at Alice Holt in the mid-sixties. This gave more consistent results and proved a severer test; so some of the ratings had to be reviewed. It was however encouraging that high degrees of resistance were confirmed in certain clones of *Populus trichocarpa*, which (with its hybrids) is silviculturally the most useful poplar for British conditions.

Cankers due to *Phomopsis* on Douglas fir and occasionally Japanese larch, which had attracted a good deal of attention before the war, have given rise to very little anxiety since. A disease (whilst not new) which has assumed greater importance in the post-war period, is the rust *Peridermium pini*, which causes very destructive lesions on stems and branches of Scots pine. It is commonest in north-east Scotland, but can be seen nowadays in East Anglia. It has been studied recently by J. S. Murray, C. S. Millar and B. J. van der Kamp (1969) of the Department of Forestry, Aberdeen, with Commission support. *Peridermium* does not require an alternate host and it was shown that it is able to enter shoots through the stomata of healthy needles, and also through bark lesions. An interesting finding by Van der Kamp was that there is some measure of control of the rust by a fungal hyperparasite *Tuberculina maxima*, which lives in *Peridermium* lesions and appears to inhibit spore production.

Another disease causing lesions on pine stems, chiefly on Corsican pine, is associated with the fungus *Crumenula sororia*, and though this is not such a serious matter as *Peridermium*, there is some suggestion that it is on the increase also. *Crumenula* was found by S. Batko and R. G. Pawsey (1964) on some of the southern heaths in the early sixties where its cankers seemed to be a secondary complaint on poorly growing pines. Recently however it has turned up in Fife on quite vigorous thirty-year-old Corsican pine, and it has also been recorded on Lodgepole pine in Scotland. The biology of this fungus has been studied very recently by A. Manap Ahmed and A. J. Hayes (1971) of Edinburgh University, and it seems likely that it has a degree of pathogenicity. As the incidence of cankers is highest on the north and north-east sides of the stems, and infection worst on ground with northern aspect, it is probable that there is some connection with temperature.

The Blister rust of five-needled pines, *Cronartium ribicola*, had discouraged the planting of Weymouth pine *Pinus strobus* (one of our earliest exotics), before the Forestry Commission's time. Unlike *Peridermium*, *Cronartium* requires an alternate host (*Ribes* species), and it was thought worthwhile to find out whether *Pinus strobus* could in fact be grown on sites at least a mile distant from currant bushes. A number of trial plantations were put down, and it was noticed that many of the cankers observed had arisen from infection in the nursery. Given clean stocks to start with, the suggestion is that *Pinus strobus* could in fact be grown on suitably isolated sites. Scion material of American selections for resistance was imported also, and exposed to infection from currants suffering from rust, along with a number of other five-needled pines. There are clearly possibilities in breeding for resistance in *P. strobus* or hybridisation in five-needled pines using the Asiatic species which are little affected, but it is doubtful whether there is a strong case for the five-needled pines at the present time.

An alarming disease of sycamore—the Sooty bark disease—turned up in the environs of London just after the war, and the preliminary surveys suggested that it might prove serious. The fungus *Cryptostroma corticale* was found associated with the trouble, and J. A. Townrow (1954) and N. F. Robertson of the Botany School of Cambridge investigated the biology of this fungus, which appeared to have a definite but not high degree of pathogenecity. Curiously, after the original flare-up, the disease declined steadily, though one or two cases have been recorded in other parts of the country. Bark disease in beech, a common complaint in old trees, was investigated, but the complex of the beech *Coccus* insect, the *Nectria* fungus, and (probably) soil factors has not been disentangled.

Though we have a fair list of specific foliar diseases, none is perhaps of the first importance economically. Some are primarily nursery complaints. *Didymascella* (*Keithia*) has long been troublesome in the raising of *Thuja plicata*. R. G. Pawsey (1960), whilst at Nottingham University, studied the biology of the fungus, and showed that the carry-over of infection from one season to the next was largely in the plant, rather than on the soil. One practical measure of avoidance was to rotate *Thuja* seedbeds and lines through a number of selected nurseries free from *Thuja* hedges or other obvious sources of infection. This worked tolerably well for the Commission but it is not helpful to the private nurseryman. One interesting line discovered fortuitously by the Geneticist was to raise stocks from cuttings from mature trees; such material being physiologically too old to be readily infected. However, search was made for an effective fungicide. Sprays based on cycloheximide gave the best results, though the substance is unpleasantly toxic. However, approved formulations have recently been introduced. *Meria* on larch used to be a source of loss, and some work was done on sprays, colloidal sulphur giving a good measure of control. With improved nursery technique, however, the turnover of stocks became much faster, and the build-up of infections from the litter was greatly lessened. *Lophodermium* on pines is not usually serious in this country, and as an occasional nursery disease it is easily avoided by siting nurseries a respectable distance from pines. The needle-cast *Rhabdocline*, which is destructive on continental provenances of Douglas fir, has attracted some attention by straying onto oceanic provenances, but it is doubtful whether the effects are going to be serious.

One of the few new records in the period was the Brown spot disease of pines, *Dothistroma pini* (Murray and Batko, 1962), which appeared at Wareham Nursery, Dorset, in 1957. This is well known in America, but of most importance on *Pinus radiata* in summer-rainfall regions of East Africa, where its effects are disastrous. It shows no signs of establishing itself here.

The pathology of young seedlings has been an important ancillary investigation to the main nursery nutrition programme and is mentioned in Chapter 13. *Botrytis* has been the commonest cause of loss in the older seedbed, and has usually been connected with early autumn frosts. It has proved extremely difficult to control by fungicides, and the avoidance of frost-susceptible species and provenances in very frost-prone nurseries is the obvious precaution.

A new pathological interest is in disease due to virus. Till recently there were few authentic cases in the literature on forest trees; no doubt (as Peace would have said) partly explained by the shortage of virologists! The Commonwealth Forestry Institute started work in the mid-sixties on virus diseases of insect pests and also on virus diseases of trees. In the latter field the work has been largely exploratory, but useful results were obtained in a study of Poplar mosaic virus by P. G. Biddle and T. W. Tinsley (1967–72). It was found that a number of the standard clones were infected, and that the virus was capable of reducing growth in the nursery to an important extent; following this, nursery stocks were rogued for virus incidence. Fortunately, however, the virus did not seem capable of significantly reducing growth of older poplars in plantation. The work has shifted towards the conifers.

The British climate being what it is, few seasons go by without observable damage from one or other vicissitude, and only those forms which have attracted considerable amounts of work need be mentioned. Low temperature injuries have been common enough, particularly in the winter of 1962/3 (when there were a number of records of bark damage and

stem crack), but nowhere catastrophic, and there has been nothing comparable with the May frost of 1935. One special aspect of frost which has been studied is the micro-climate of East Anglian pine regeneration fellings, where it has been shown that conditions in clear felled areas surrounded by forest are, if anything, more frosty than the original open Breckland, and the effects are accentuated by the insulation of the ground by felling slash. Drought has been of chief interest in connection with stem crack; the hot dry summer of 1959 in particular bringing in many records of this form of damage. It is most common in *Abies procera* and *Abies grandis*, and since the status of these two trees has been in question, special investigations have been carried out recently on this failing (Greig, 1969). The incidence of stem crack in *Abies procera* under British conditions does appear to be unacceptably high. In *Abies grandis* however the incidence is lower, and analyses of the cracks themselves together with experimental conversions of parcels exhibiting stem-crack had shown that the potential loss from this fault is not as high as might be expected, and on a countrywide basis may well be insignificant.

Exposure effects have been too varied and numerous for detailed study, and the most profitable approach is likely to be that of the physiologist. Atmospheric pollution is however a topic for the pathologist, and several cases have been investigated. Peace made observations on suspected fluorine damage to trees from aluminium works near Fort William in 1948, and later investigated an interesting case of fluorine damage from brick-work fumes, which in this particular instance were concentrated near the ground by an exceptionally intense inversion. Sulphur dioxide, which is the common pollutant of industrial areas, is so general in some districts as almost to rank as a climatic factor, and indeed it has been very difficult to separate its effects from those of exposure where the two are combined. This was particularly the case in Line's work in the Pennines of Lancashire and the West Riding of Yorkshire.

Forest Entomology (*Plate 66*).

H. S. Hanson, the first holder of the post of Forest Entomologist at Alice Holt, had in fact worked for the Commission since 1941, and had undertaken extensive surveys of war felling areas in connection with the risks to new plantations from insect pests, especially *Hylobius* and *Hylastes* (Hanson, 1943). Dr W. B. R. Laidlaw also worked for the Commission for a short time in the immediate post-war period, specialising on aphids and adelgids in Scottish forests (Laidlaw, 1930; 1934). Hanson was succeeded by Dr Myles Crooke in 1952, who left to join the staff at Aberdeen University in 1960; D. Bevan, who had joined the Section in 1949, took over the post from

Myles Crooke and has continued to lead the work ever since.

The extensive war fellings in coniferous woodlands had of course provided ample breeding material for *Hylobius* and *Hylastes*. The policy at this time was to wait till the stumps had ceased to provide suitable breeding sites before replanting; this was satisfactory enough so far as damage from weevils and beetles was concerned, but wasted time and sacrificed favourable planting conditions. The protection of the young plant itself was a later development. An indirect hazard of the war was the import of un-barked softwood logs (mainly spruce) from Germany in the period shortly after the end of hostilities in Europe. Some consignments brought with them thriving broods of bark-beetles, of which *Ips typographus* was regarded as the least welcome immigrant (Laidlaw, 1947; Forestry Commission, 1948). The diversion of the most dangerous consignments to city mills and away from rural ones was an effective control measure, and fortunately this episode did not result in the establishment of an important pest; the conditions were of course against it as the acreage of mature spruce in the country was by this time very small.

Most of the Coleopterous insects which breed under bark and do feeding damage of various kinds are well-known creatures to the entomologist, and become important only when breeding sites multiply from natural causes or unsound management. There have been, for instance, temporary outbreaks of the pine-shoot beetle *Tomicus piniperda* (*Myelophilus*) in the early fifties where the conversion of poles from thinnings had been centralised at depots, but this has been readily controlled by rationalising the conversion process to avoid large dumps of unbarked material during the breeding period. Much more important was the gale of January 1953, which blew down some forty million cubic feet of timber, mainly pine, in the north-east of Scotland (Crooke, 1955). (Plates 62 and 63.)

Tomicus was the obvious hazard in this occurrence, but as Myles Crooke pointed out, there were some redeeming features. Most of the trees thrown intact remained too moist for successful breeding in the first season after the blow. An interesting opportunity for control was provided by the many snapped trees which attracted most of the original breeding population, and had it been possible to treat them as such, would have served excellently as traps. Surveys of the breeding possibilities for *Hylobius* and *Hylastes* suggested that the expected build-up would be relatively short in duration because of the greater rate of drying out of windthrown stumps, compared with stumps from felled trees. Following this gale, some of the first successful experiments were carried out on the control of *Tomicus* in stacked pine timber

using the modern insecticide DDT. Following recent work, treatments for the control of *Tomicus* in stacks are now available using BHC preparations applied either through a mist blower or a high-volume machine. These methods are regarded as emergency measures in case of dislocations in the transport of logs from the forest (Bevan, 1962 b).

Reverting to *Hylobius* and *Hylastes*; the beginning of regeneration fellings was an incentive to new work on these familiar pests. *Hylobius* being readily attracted to traps (usually small conifer billets covered with fresh bark), the use of the new persistent insecticides suggested itself as a way of dispensing with hand collection of weevils from the traps (Hanson, 1952 b). This in fact showed some promise, but trapping still meant a good deal of work. The most attractive line was to put the insecticide on the plant itself. The first applications were dusts of Lindane, DDT etc. in suitable carriers, and these gave good protection when applied in plants in situ. The advantages of treatment in mass were obvious, and led to experiments in dipping plants in insecticidal solutions (Crooke, 1957). Satisfactory treatments of this kind became available in the late fifties, but the dips occasionally showed some phytotoxicity and safer formulations were sought. The early dips had treated the tops of the plants only, but later experiments included dipping the whole plant to extend the protection (primarily against *Hylobius*) against *Hylastes* also, which tends to feed lower down (Stoakley, 1968). Solutions of water-based formulations containing Lindane have proved very effective, and of low phytotoxicity, and by now are standard treatments on sites where *Hylobius* and *Hylastes* damage is to be expected. This is an interesting application of the persistent insecticide, and as the quantities are small and the application is localised, there is not much danger of getting undesirable amounts into the cycle.

On the whole British experience with the bark beetles has been fortunate. The scare over *Ips typographus* has been mentioned; a greater hazard is probably *Dendroctonus micans* which has been particularly damaging to Sitka spruce on the Continent, and appears to behave nearly, if not quite, as a primary pest. J. M. B. Brown and D. Bevan (1966) made a special study of this beetle in Europe in 1964. The greatest danger is that the beetle might be introduced accidentally and take hold in plantations of Sitka spruce on poor sites, especially those which are too dry for the tree. One European bark beetle new to Britain, *Ips cembrae*, does seem to have established itself during the period. First observed in 1955, it is now found over a considerable area of north-east Scotland (Crooke and Bevan, 1957; Crooke and Kirkland, 1960).

Wood-boring insects in this country are usually

troubles of stored or converted timber rather than forest pests. Pin-hole borers (ambrosia beetles) have however been the cause of considerable degrade of spruce logs lying in the forest in the west of Scotland on one or two occasions. The damage became obvious in the late fifties in logs arriving at the Cowal-Ari mill at Strachur, Argyll, and was found to be due in the main to *Trypodendron lineatum*. Investigations into the biology (Balfour and Paramonov, 1962) of these creatures showed that early winter-felled logs were very much more attractive than spring-felled logs, and as the pin-hole borers began to attack logs at the end of April, it was obvious that management should get the logs off the ground before this, and give priority to the logs felled earliest (Bevan, 1962a; Bletchly, 1960). At the same time experiments were done on protection of stacks, and it was found that Lindane gave good control; a useful measure in case of emergency. It was also noted that stumps were not a very important reservoir for pin-hole borers (Balfour and Kirkland, 1963).

Of other wood borers, the wood wasps (*Sirex* species) are definitely ranked as forest pests, since their larvae work in standing trees (Hanson, 1939). They are of little importance here, and the chief interest in them has been at the Commonwealth level, for they have been disastrous introductions to New Zealand and, latterly, Australia, and there has been much interest in their parasites for biological control. Hanson sent the parasite *Ibalia* to New Zealand in 1950.

Primary defoliators, having the greatest potential for damage in coniferous forest, have naturally attracted a great deal of attention from the entomologists. A good deal of this work has been concerned with the biology of the likeliest candidates to rank as major pests. One group to which Hanson gave much attention is that of the sawflies, which includes several species capable of doing substantial damage to larch, and others which feed on spruces and pines. Though there have been numerous occasions when local epidemics of sawflies have appeared very unpleasant, they seem very prone to population collapses. Parasitism is important in the sawflies, and parasites have been bred and released on one or two occasions; but virus diseases are a greater cause of mortality. Crooke noted a polyhedral disease in the Large larch sawfly (*Pristophora erichsoni*) in 1953, and outbreaks of one of the pine sawflies (*Neodiprion sertifer*) usually end this way; it has been found that a weak suspension of virus-killed larvae sprayed on infested pines is a very efficient form of biological control (Rivers and Crooke, 1960). On the whole, the sawflies are not now regarded as pests of the first importance.

Most of the potentially serious defoliators belong to the *Lepidoptera* and several of the moths which are responsible for epidemics on the Continent are indigenous to Britain. In European pine forests, *Bupalus piniarius*, *Lymantria monarcha*, and *Panolis flammea* rank as serious pests, and their behaviour in large-scale pine monocultures here has been carefully watched. So far only the first has, on a few occasions, increased to epidemic proportions.

Populations of *Bupalus piniarius*, the Pine looper moth, large enough to defoliate pines, first appeared in Cannock Chase, Staffordshire, and the dune plantations of Culbin, Laigh of Moray Forest, Moray, in 1953, and surveys by pupal counts in these forests suggested that considerable areas were at risk of defoliation in the following season. It was hence decided to resort to aerial spraying in the summer of 1954—the first exercise of this kind to be undertaken in Britain. Some 2,500 acres were sprayed at Cannock and 3,500 at Culbin using one lb of DDT per acre (Crooke, 1959). The operation was entirely successful as a short-term measure, further damage being averted, and the population of *Bupalas* settling down to an endemic level. (Plates 64, 65, 67.)

Such emergency measures were not regarded as a satisfactory solution to the *Bupalus* problem (or indeed, to any insect epidemic), and *Bupalus* has remained a major project on the Entomologist's programme over the rest of the period. From Bevan's own work and that of N. W. Hussey at Edinburgh University a great deal is now known about the biology of *Bupalus* in Britain, its feeding habits, fecundity, and general population dynamics (Bevan, 1955; 1961; 1966a; Bevan and Brown, 1961; Bevan and Paramonov, 1957; 1962; Hussey, 1956; 1957). Joan Davies has studied its parasites, one of which at least (*Cratichneumon nigritarius*) appears to exercise significant control in certain seasons (Davies, 1957). Environmental studies have failed to give any simple explanation why certain forests are prone to epidemics, whilst others (such as Thetford), which appear very suited to *Bupalus*, continue to have low populations of the insect. It is a pest associated with pure plantations in the pole stage and older woods. The replacement of Scots pine by Corsican within its useful range is one obvious measure, since the latter tree appears much less prone to *Bupalus* attack.

In 1963, *Bupalus* levels at Cannock Chase necessitated a second spraying operation in that forest, over some 1,400 acres. On this occasion the Nature Conservancy collaborated in the exercise to look for side-effects on wild life, and to study the persistence of residues. *Bupalus* so far remains the only defoliator which has called for such measures in British forestry. Oak is frequently defoliated by *Tortrix viridana* (Hussey, 1961) early in the season without suffering much, since it re-flushes readily. There has long been controversy over the role of birds, especially titmice, in the control of defoliators.

Dr D. Lack (1950–53) of the Edward Grey Institute, Oxford, conducted studies of tits in nesting boxes erected in the Forest of Dean and other forests adjacent to the Commission's Forest Schools, the students carrying out much of the field observations. This yielded a great deal of valuable information on the breeding rates and feeding habits of tits and one or two other birds. More specific efforts to relate tit populations to those of the defoliator *Bupalus* have been made by Myles Crooke (1964–71) at Aberdeen University since 1962 at Culbin Forest, Moray. This study is a long-term one, entailing the maintenance of a high tit population by ample provision of nesting boxes (and latterly winter feeding) on certain plots, which will be compared in respect of *Bupalus* numbers with others chosen for their initial similarity in all conditions. While no one expects tits to breed fast enough to control an epidemic, there is the possibility that they might help to damp down numbers below the epidemic level.

One of the few defoliators which have appeared on Sitka spruce is the tortricid moth *Zeiraphera diniana*, of which there was a minor epidemic at Hope, Matlock Forest, Derbyshire, in 1967. This moth was, however, subject to very high predation and parasitism. Several well-known lepidopterous insects are concerned with damage other than defoliation, such as shoot and bud tunneling. None has been of great importance in this period. The Pine-shoot moth *Rhyacionia buoliana* was taken very seriously as a pest of young pine plantations in the pre-war period (Brooks and Brown, 1946), but its very conspicuous damage in young plantations is now known to have little economic significance in British conditions (Forestry Commission, 1956).

The large and complex group of sap-sucking insects, the aphids and adelgids, contain one creature of major importance, *Elatobium* (*Neomyzaphis*) *abietinum*, and many of local or transitory significance. Another, *Adelges nordmannianae* (Busby, 1964; Varty, 1956), would have higher ranking now if it had not already won its battle, and deprived us of the use of the European Silver fir, *Abies alba*. The only recent approach to the problem of *Adelges* has been through provenance (Chapter 22). An interesting aid to the study of aphids and adelgids has been the setting up of an improved type of suction trap at Alice Holt (Stickland, 1967), which has given a great deal of information on the species present and their flight periods. A major systematic work was undertaken by C. I. Carter which culminated in the issue of Bulletin 42 on conifer woolly aphids (Adelgidae) in Britain (Carter, 1971).

Elatobium and its biology has been very thoroughly studied. Hanson (1952a) observed one of the most important points about it in the early fifties—that over-wintering populations were greatly favoured by mild weather. The most serious epidemics have always followed exceptionally "soft" winters.

W. H. Parry (1968–71; 1969) at Aberdeen, who did some very detailed work on *Elatobium* in the late sixties, confirmed that winter temperature was the main controlling factor on population. Some field officers, particularly those who have Sitka spruce growing under inadequate rainfall, take a very serious view of it, and in bad years the entomologists have frequently been asked to "do something about it". It has never been clear that there is any promising approach, and the insect is best regarded as an inseparable associate of Sitka spruce (Bevan, 1966b). It has however been thought worth while to estimate the increment loss caused by *Elatobium*, and this has been done in two ways, firstly by studying ring width in relation to known *Elatobium* epidemics in the past, and secondly by the more direct method of comparing growth in plots kept free of *Elatobium* by insecticides with that in plots left to the normal course of infestation. It is clear enough that the loss is considerable, probably, in a bad year, of the order of 20 per cent.

Seed-feeding insects are not of great importance in British forestry, with the possible exception of the seedflies, *Megastigmus* species, which breed in the seed of several conifers. The group has been studied by Hussey (1952–54; 1954; 1955) at Edinburgh University. The best known is *Megastigmus spermotrophus* on Douglas fir. Infestations from this insect cause considerable losses of seed, and with the future production of improved home seed supplies in view, it has been thought worth while to investigate the protection of cones against *Megastigmus*. J. T. Stoakley (1967) went into this in some detail, and obtained reasonably encouraging results by spraying with malathion during the period of oviposition. There was however a very strong inverse relationship between infestation levels and cone crop, which suggested that for normal requirements, special efforts to collect large quantities should be made in bumper years. But for breeding work, the possibility of protection is there. Incidentally, this project gave some very useful experience in radiography and seed testing.

Unwanted creatures of this kind may well arrive in imported seed. On two occasions in the fifties Sweet chestnuts and acorns arrived with larvae of a weevil *Curculio elephas* which is not indigenous.

Insect pests in the nursery have not been of great importance over the period; the few species which have required some research attention are mentioned in Chapter 14.

An important development in the late sixties has been the establishment of the Insect Pathology Unit at the Commonwealth Forestry Institute, Oxford. The Unit specialises in virus diseases of insects, which are often the best prospects for bio-logical control. The Institute of Tree Biology of the Natural Environment Research Council also intends to work on the topic of biological control.

Mammals and Birds

Most of the important insect pests of woodlands constitute a private headache for the forester, but in this country almost all the higher animals which are capable of damage in woodlands are equally capable of damage to agricultural and horticultural crops. There are also the predators, all of which are entirely beneficial to forestry, but not necessarily to other interests. And as forests account for a small percent-age of the land surface compared with most Euro-pean countries a forest wildlife policy must take account of other land users. A good deal of the work which has been done on animals (especially deer) has not in the first place been labelled or organised as research but has started as field-craft, and has been recognised later as a special branch of management requiring research support. Skilled field naturalists, such as J. S. R. Chard and P. F. Garthwaite amongst senior officers, have had a great deal of influence on the management of wild life in forests during this period.

The Research Branch became directly involved in a major investigation concerning mammals in 1954, when alarm about the increasing damage caused by the introduced Grey squirrel provided the incentive for a programme of research into the biology of the creature, and methods for its control. This was a joint exercise with the Pest Control Division of the Ministry of Agriculture, and the investigations were planned by Mrs Visozo (Monica Shorten), the recognised authority on squirrels. Primary studies were concerned with the population dynamics of squirrels, and in areas set aside for this, numbers, movements and reproduction rates were estimated by catch-mark-release methods, and information built up on breeding rates in relation to food supplies, mast years in beech and oak being the most signifi-cant events (Shorten and Courtier, 1955). The main damage period, when bark is stripped from broad-leaved trees, particularly beech, sycamore and ash, appeared to fall in a period of scarcity of food between the spring, when buds and flowers are eaten, and the late summer and autumn, when fruits begin to be available. Later research suggested that the social organisation of the Grey squirrel was also important in stimulating the occurrence of damage.

To encourage the destruction of the Grey squirrel, a bounty on tails was paid for some years. This is not usually successful with small fast-breeding mammals, and general population studies on the Grey squirrel showed that it was no exception; it was in fact apparent from an early stage that the squirrel was

here to stay, and fluctuations in numbers (which were large) had more to do with food supplies than with the numbers of tails paid for. Hence this measure was dropped; and the later investigations have assumed that the creature would never be eradicated, but that control measures might avert damage to vulnerable woods. A promising line followed by Judith Rowe (1967) was intensive cage trapping in such vulnerable areas (hardwood pole stands etc.) in the period May/July, preceding and embracing the main damage season. In trial areas, this succeeded in averting damage for six successive seasons. Comparisons were made between the two main methods of destroying squirrels; shooting in combination with drey poking, and trapping. Trap-ping has always appeared the better method. Various forms of live cage traps and humane killing traps were investigated. K. D. Taylor (1963) and H. G. Lloyd of the Ministry of Agriculture began enquiries into the use of the rodenticide warfarin in the late fifties. Because of differences in the wording of legislation concerning poisons and vermin, it was possible to conduct field trials in Scotland but not (then) in England. Practicable and safe baiting methods were worked out, and wherever the method is regarded as legitimate, warfarin used locally to reduce numbers and protect valuable stands may well prove one of the answers. (Regulations to permit the use of warfarin over most of England and Wales were promulgated in June, 1973).

Research work on deer has been mainly directed to furthering the policy of managing deer popula-tions resident in the forest at a level where damage to woodlands is acceptable. The field-craft of the ranger is of the first importance, but research comes in on the essential biological detail; methods of determin-ation, techniques of recording movement, marking, counting etc. Red deer are traditionally excluded from the forest, but in Galloway one isolated herd lives in the forest and has been the subject of special study. This has included the study of vegetation for winter feed. In all this work there is common ground with the Nature Conservancy.

Fencing for the exclusion of deer has attracted a good deal of attention recently after a long period of traditional designs. The main advance stems from the adoption of high tensile spring steel wire, which cuts the number of posts. The use of light polythene netting for the upper (specifically deer) part of the fence was effective but was discontinued because animals tended to become entangled. There has been some success with electric fencing, but this entails fairly frequent applications of herbicides to ground vegetation to avoid short-circuiting, and these high maintenance costs make it impractical for forest use.

A number of commercial repellant substances have been tried. Some of these will give protection up to about twelve weeks, which is not much good in forestry, but may be useful in arboriculture or gardening.

Amongst other mammals, some work has been done on the voles. Baiting techniques and densities for the use of warfarin have been worked out, and in epidemics it seems likely that this would prove useful in specially valuable crops such as experimental plantations. J. D. Lockie (1967) of Edinburgh University studied voles and their predators, weasels and stoats, in the mid-sixties, and obtained most interesting information on the social organisation of these animals. The forester's best defence against voles is probably to get his crops into canopy as quickly as possible. One interesting but localised new rodent, the Edible dormouse (*Glis glis*), attracted some attention in the sixties by doing squirrel-like damage to trees, mainly conifers, at Wendover, Chiltern Forest, Buckinghamshire.

Birds have often done damage to seedbeds in nurseries. Some attention has been given to repellant chemicals as seed dressings, but where this form of damage is common, it has been found more profitable to use modern synthetic fibre netting. The only other form of bird damage of any significance in British forestry is caused by massed starling roosts in plantations (Bevan, 1962c; Elgy, 1972). After a few seasons nitrogen levels from droppings may become toxic. The technique is to break up the roosts and "move them on" (like Jo in *Bleak House*). Combinations of bangers and amplified distress calls have proved effective. (No thought appears to have been given to leaving them alone and setting up local guano industries.)

The excellent national cover of foresters which now exists has made the use of well-designed questionnaires a valuable method of getting information on the distribution of animals and birds in forest areas, and the incidence of damage. This is of course of quite general significance, and not only to forest protection.

Forest Fires

Much of the system of fire protection as practised in this country is based on methods developed by older forest authorities, modified and further developed to suit British conditions. Our main feature is the presence of several highly flammable ground vegetations, especially in spring, which with our windy climate, make for rapid spread of ground fires. Owing to our rather high humidities, crown fires in older crops are fortunately rare. Hence in setting standards for prevention and suppression much of the effort has gone to speed in firespotting, communications (such as use of radio), access, etc.

It is a difficulty that with some of the dominant vegetation types in our changeable climate, hazard

may change from day to day; hence though useful work has been done on this (notably by G. D. Rouse), hazard ratings based on the humidity of fuels are not so easy to apply as in countries with more settled weather conditions.

The Research Branch has been brought in on certain points. Firstly, the appearance of new herbicides in forest practice in the late forties and early fifties suggested their application to fire traces in place of mechanical or hand cultivation. G. D. Holmes tested a number of substances for this purpose whilst working on herbicides generally (Wood and Holmes, 1957). By 1960 it appeared that none of the substances then available was cheaper than mechanical cultivation, but several non-persistent and persistent herbicides might have their use where cultivation was not possible. The position was changed by the arrival of paraquat, since breaks unsuited to cultivation or mowing can be kept weed free fairly cheaply by back-pack spraying. This "dessicant" also allows control burning to be done on marches (or even inside plantations) with greater safety, since grasses (especially *Molinia*) can be rendered flammable whilst the adjacent vegetation is in full summer growth.

Holmes and D. A. Cousins also collaborated with C. A. Connell on fire retardants (Holmes and Fourt, 1961; Connell and Cousins, 1969). There were two approaches. Firstly, several known retardants such as borates, bentonite clay suspensions, and ammonium phosphate solutions were tried out on typical forest fuels, with some success. There were however some disadvantages connected with such retardants, as they usually required special equipment to apply. The second approach was to make a given quantity of water go further, by increasing its viscosity. Sodium alginate proved very satisfactory for this purpose, and has the advantage of being introduced easily into the normal fire-fighting equipment, pumps, back-pack sprays etc. The use of "viscous" water helps enormously in giving safe margins for control burning and counter firing in actual fighting operations, besides its value at the fire front itself.

Basic research on forest fuels and the rate of spread of fires has been carried out by P. H. Thomas (1962–70) and his colleagues at Boreham Wood (Department of the Environment and Fire Offices' Committee Joint Fire Research Organisation) since 1961. In this work the laboratory side has consisted of the detailed study of fires in wooden cribs designed to simulate forest fuels. The characteristics of the fires have been studied in still air and with winds of known velocities, and the part played by radioactive and convective transfers of heat determined for different circumstances of fuel and wind. From such models, mathematical formulae have been deduced relating rates of spread to the main physical factors of the fuel and to wind speed. These have been tested in experimental "burns" in natural conditions, with a very fair measure of agreement, though the heterogeneity of natural conditions, and especially the movement of burning brands makes it difficult to construct expressions in purely physical terms. There may be two practical applications from this kind of work. Firstly, it should have some bearing on the width of breaks in young plantations in relation to the lengths and angles of flames which can be expected under the worst conditions. Secondly, it should provide a basis for the spacing of firebreaks and accesses in terms of the likely rate of spread of fire at the most hazardous stage in the life of the plantations. It may be that experience has already given good approximations, but it might well be worth re-examining the layouts in some of the most dangerous forests on theoretical grounds.

REFERENCES

AHMED, A. MANAP, and HAYES, A. J. (1971). *Crumenula sororia* Karst. associated with cankering and die back of Corsican pine. *Scott. For.* 25 (3), 185–200.

BALFOUR, R. M., and KIRKLAND, R. C. (1963). The effect of creosote on populations of *Trypodendron lineatum* breeding in stumps. *Rep. Forest Res., Lond.* 1962, 163–166.

BALFOUR, R. M., and PARAMONOV, A. (1962). Is the flight of *Trypodendron lineatum* (Oliv.) (Col. Scolytidae) strictly necessary ? *Ent. mon. Mag.* 98 (1174/1176), 66–67.

BATKO, S., and PAWSEY, R. G. (1964). Stem canker of pine caused by *Crumenula sororia*. *Trans. Br. mycol. Soc.* 47, 257–261.

BEVAN, D. (1955). The status of the Pine looper moth (*Bupalus piniarius* L.) in Britain in 1953. *Rep. Forest Res., Lond.* 1954, 158–163.

BEVAN, D. (1961). Insecticidal control of the Pine looper in Great Britain. II. Population assessment and fogging. *Forestry* 34 (1), 14–24.

BEVAN, D. (1962a). The Ambrosia beetle or Pinhole borer *Trypodendron lineatum* Ol. *Scott. For.* 16 (2), 94–99.

BEVAN, D. (1962b). *Pine-shoot beetles.* Leafl. For. Commn 3. Revised.

BEVAN, D. (1962c). Starling roosts in woodlands. *Q. Jl For.* 56, 59–62.

BEVAN, D. (1966a). *Pine looper moth, Bupalus piniarius.* Leafl. For. Commn 32.

BEVAN, D. (1966b). The Green spruce aphis, *Elatobium* (*Neomyzaphis*) *abietinum* Walker. *Scott. For.* 20 (3), 193–201.

BEVAN, D., and BROWN, R. M. (1961). The Pine looper moth *Bupalus piniarius* in Rendlesham and Sherwood Forests—1959. *Rep. Forest Res., Lond.* 1960, 172–179.

BEVAN, D., DAVIES, J. M., BROWN, R. M., and CROOKE, MYLES (1957–1962). The Pine looper, *Bupalus piniarius. Rep. Forest Res., Lond.* 1957, 68–72; 1958, 72–74; 1959, 70–71; 1960, 63–66; 1961, 60–61 and 176–182.

BEVAN, D., and PARAMONOV, A. (1957). Fecundity of *Bupalus piniarius* in Britain, 1955. *Rep. Forest Res., Lond.* 1956, 155–162.

BEVAN, D., and PARAMONOV, A. (1962). Fecundity of the Pine looper moth, *Bupalus piniarius. Rep. Forest Res., Lond.* 1961, 174–176.

BIDDLE, P. G., and TINSLEY, T. W. (1967–1972). Virus diseases of forest trees. *Rep. Forest Res., Lond.* 1967, 156–159; 1968, 149–151; 1969, 150; 1970, 179–180; 1971, 134; 1972, 153.

BLETCHLY, J. D. (1960). A review of factors affecting Ambrosia beetle attack in trees and felled logs. Paper for 11th *Int. Congr. Entomology*, Vienna.

BROOKS, C. C., and BROWN, J. M. B. (1946). *Studies on the Pine shoot moth (Evetria buoliana* Schiff.). Bull. For. Commn, Lond. 16.

BROWN, J. M. B., and BEVAN, D. (1966). *The Great spruce bark beetle.* Bull. For. Commn, Lond. 38.

BURDEKIN, D. A. (1972). Bacterial canker of poplar. *Ann. appl. Biol.* 59, 295–299.

BUSBY, R. J. N. (1964). A new adelgid on *Abies grandis* causing compression wood. *Q. Jl For.* 58 (2), 160–162.

CARTER C. I. (1971). *Conifer woolly aphids (Adelgidae) in Britain.* Bull. For. Commn, Lond. 42.

CONNELL, C. A., and COUSINS, D. A. (1969). Practical developments in the use of chemicals for forest fire control. *Forestry* 42 (2), 119–132.

CROOKE, M. (1955). Forest insects in the gale-damaged woodlands of north-east Scotland, 1953–1954. *Rep. Forest Res., Lond.* 1954, 163–169.

CROOKE, M. (1957). Experiments on the control of the Pine weevil *Hylobius abietis* L. *Proc. 7th Br. Commonw. For. Conf.*, Australia and New Zealand.

CROOKE, M. (1959). Insecticidal control of the Pine looper in Great Britain. I. Aerial spraying. *Forestry* 32 (2), 166–196.

CROOKE, M. (1964–1971). Studies on tit and Pine looper moth populations at Culbin Forest. *Rep. Forest Res., Lond.* 1963, 122–124; 1964, 133–134; 1965, 190–200; 1967, 165; 1968, 159; 1969, 155; 1970, 185; 1971, 138–139.

Crooke, M., and BEVAN, D. (1957). Note on the first British occurrence of *Ips cembrae* Heer (Col. Scolytidae). *Forestry* 30 (1), 21–28.

CROOKE, M., and KIRKLAND, R. C. (1960). Resurvey of distribution of the Bark beetle *Ips cembrae*. *Rep. Forest Res., Lond.* 1959, 167–169.

DAVIES, J. M. (1957). Parasites of *Bupalus piniarus. Rep. Forest Res., Lond.* 71–72.

DAY, W. R. (1932). Defoliation of larch as a cause of disease. *Forestry* 6, 113–124.

DAY, W. R. (1950). Frost as a cause of die-back and canker of Japanese larch. *Q. Jl For.* **44** (2), 78–82.

DAY, W. R., and SANZEN-BAKER, R. G. (1939). *Preliminary report on the investigation into the afforestation of chalk downland.* Typescript 19 pp. For. Commn, Lond.

ELGY, D. (1972). Starling roost dispersal in forests. *Q. Jl For.* **66** (3), 224–229.

FORESTRY COMMISSION (1948). *The Spruce bark beetle.* Leafl. For. Commn 26.

FORESTRY COMMISSION (1952). *Sooty bark disease of sycamore.* Leafl. For. Commn 30.

FORESTRY COMMISSION (1956). *The Pine shoot moth.* Leafl. For. Commn 40.

FORESTRY COMMISSION (1967). *Keithia disease of Western red cedar, Thuja plicata.* Leafl. For. Commn 43.

FORESTRY COMMISSION (1970). *Fomes annosus.* Leafl. For. Commn 5.

GIBBS, J. N. (1967a). A study of the epiphytic growth habit of *Fomes annosus. Ann. Bot.* **31**, 755–774.

GIBBS, J. N. (1967b). The role of host vigour on the susceptibility of pines to *Fomes annosus. Ann. Bot.* **31**, 801–815.

GIBBS, J. N. (1968). Resin and the resistance of conifers to *Fomes annosus. Ann Bot.* **32**, 649–665.

GLADMAN, R. J., and GREIG, B. J. W. (1965). *Principal butt rots of conifers.* Bookl. For. Commn 13.

GREIG, B. J. W. (1962). *Fomes annosus* (Fr.) Cke. and other root-rotting fungi on ex-hardwood sites. *Forestry* **35**, 164–182.

GREIG, B. J. W. (1969). On drought crack in Grand fir. *Suppl. Timb. Trades J.* March, 26–27.

GREIG, B. J. W., and BURDEKIN, D. A. (1970). Control and eradication of *Fomes annosus* in Great Britain. *Proc. 3rd IUFRO Conf. on F. annosus, Denmark,* 21–32.

GREMMEN, J. (1968). Bijdrage tot de biologie van *Brunchorstia pinea* (Karst.) Hohn., de oorzaak van het taksterven bij Oostenrijke en Corsicaanse den (Contribution to the biology of *Brunchorstia pinea* (Karst.) Hohn., the cause of a die-back in Austrian and Corsican pine). *Ned. BoschbTijdschr.* **40**, 221–231.

HANSON, H. S. (1939). Ecological notes on the Sirex wood wasps and their parasites. *Bull. ent. Res.* **30** (1), 27–76.

HANSON, H. S. (1943). The control of bark beetles and weevils in coniferous forests in Britain. *Scott. For. J.* **57**, 19–45.

HANSON, H. S. (1952a). The Green spruce aphis, *Neomyzaphis abietina* Walker. *Rep. Forest Res., Lond.* 1951, 98–104.

HANSON, H. S. (1952b). Insecticide experimental work. *Rep. Forest Res., Lond.* 1951, 107.

HOLMES, G. D., and FOURT, D. F. (1961). The use of herbicides for controlling vegetation in forest fire breaks and uncropped land. *Rep. Forest Res., Lond.* 1960, 119–137.

HOLTAM, B. W. (1968). *Blue stain of coniferous wood.* Leafl. For. Commn 53.

HUSSEY, N. W. (1952). A contribution to the bionomics of the Green spruce aphid (*Neomyzaphis abietina* Walker). *Scott. For.* **6** (4), 121–130.

HUSSEY, N. W. (1952–1954). *Megastigmus* insects attacking conifer seed. *Rep. Forest Res., Lond.* 1951, 137–138; 1952, 127–128; 1953, 120–121.

HUSSEY, N. W. (1954). *Megastigmus flies attacking conifer seed.* Leafl. For. Commn 8. Revised.

HUSSEY, N. W. (1955). The life histories of *Megastigmus spermotrophus* Wachtl (Hymenoptera: Chalcidoidea) and its principal parasite with descriptions of the development stages. *Trans. R. ent. Soc. Lond.* **106** (2), 133–151.

HUSSEY, N. W. (1956). Bioclimatic studies on the Pine looper moth. *Rep. Forest Res., Lond.* 1955, 80–83.

HUSSEY, N. W. (1957). Effects of the physical environment on the development of the Pine looper, *Bupalus piniarius. Rep. Forest Res., Lond.* 1957, 111–128.

HUSSEY, N. W. (1961). *Oak leaf roller moth.* Leafl. For. Commn 10. Revised.

LACK, D. (1950–1953). Forest ornithological investigations (The nesting of titmice in boxes). *Rep. Forest Res., Lond.* 1949, 79–80; 1950, 125–126; 1951, 138–139; 1952, 128–129.

LAIDLAW, W. B. R. (1930). A note on *Chermes cooleyi* and some of its enemies. *Scott. For. J.* **44**, 92–93.

LAIDLAW, W. B. R. (1934). Notes and observations on the conifer root aphis *Pachypapella piceae*, Htg (*Pemphigus piceae*, Tullgren). *Scott. For. J.* **48**, 79–84.

LAIDLAW, W. B. R. (1947). On the appearance of the bark beetle *Ips typographus* in Britain on imported timber with notes on preventive and control measures. *Forestry* **20**, 52–56.

LOCKIE, J. D. (1967). Voles and their predators. *Rep. Forest Res., Lond.* 1966, 111.

LOW, J. D., and GLADMAN, R. J. (1960). *Fomes annosus in Great Britain. An assessment of the situation in 1958.* Forest Rec., Lond. 41.

MANNERS, J. G. (1953; 1957). Studies on larch canker. I. The taxonomy and biology of *Trichoscyphella willkommii* (Hart.) Nannf. and related species. *Trans. Br. mycol. Soc.* **36**, 362–374 (1953). II. The incidence and anatomy of cankers produced experimentally either by inoculation or by freezing. *Trans. Br. mycol. Soc.* **40**, 500–508 (1957).

MEREDITH, D. S. (1959). The infection of pine stumps by *Fomes annosus* and other fungi. *Ann. Bot.* **23**, 455–476.

MEREDITH, D. S. (1960). Further observations on fungi inhabiting pine stumps. *Ann. Bot.* **24,** 63–78.

MURRAY, J. S., and BATKO, S. (1962). *Dothistroma pini* Hulbary: A new disease on pine in Britain. *Forestry* **35,** 57–65.

MURRAY, J. S., MILLAR, C. S., and VAN DER KAMP, B. J. (1969). Incidence and importance of *Peridermium pini* (Pers.) Lev. in north-east Scotland. *Forestry* **42,** 165–184.

MURRAY, J. S., and YOUNG, C. W. T. (1961). *Group dying of conifers.* Forest Rec., Lond. 46.

PARRY, W. H. (1968–1971). Research on the Green spruce aphid, *Elatobium abietinum. Rep. Forest Res., Lond.* 1968, 160–161; 1969, 152–154; 1970, 182–184; 1971, 136–137.

PARRY, W. H. (1969). A study of the relationship between defoliation of Sitka spruce and population levels of *Elatobium abietinum* (Walker). *Forestry* **42** (1), 69–82.

PAWSEY, R. G. (1960). An investigation into Keithia disease of *Thuja plicata. Forestry* **33,** 174–186.

PAWSEY, R. G. (1969). A reappraisal of canker and dieback of European larch. *Forestry* **42,** 145–164.

PAWSEY, R. G., and GLADMAN, R. J. (1965). *Decay in standing conifers developing from extraction damage.* Forest Rec., Lond. 54.

PEACE, T. R. (1960). *The status and development of elm disease in Britain.* Bull. For. Commn, Lond. 33.

PEACE, T. R. (1962). *Pathology of trees and shrubs with special reference to Britain.* Oxford: Clarendon Press.

PEPPER, H. W., and TEE, L. A. (1972). *Forest fencing.* Forest Rec., Lond. 80.

PHILLIPS, D. H. (1963). Forest pathology: Death and decay caused by *Fomes annosus* (Fr.) Cooke. *Rep. Forest Res., Lond.* 1962, 61–62.

PHILLIPS, D. H., and GREIG, B. J. W. (1970). Some chemicals to prevent stump colonisation by *Fomes annosus* (Fr.) Cooke. *Ann. appl. Biol.* **66,** 441–452.

PLATT, F. B. W., and ROWE, J. J. (1964). Damage by the edible dormouse (*Glis glis* L.) at Wendover Forest (Chilterns). *Q. Jl For.* **58,** 228–233.

PUNTER, D. (1962–1964). The effects of stump treatments on fungal colonisation of conifer stumps. *Rep. Forest Res., Lond.* 1961, 115; 1962, 127–128; 1963, 121–122.

RATCLIFFE, P. R. (1970a). A method of preserving animal skins. *Deer* **2** (3), 574.

RATCLIFFE, P. R. (1970b). The occurrence of vestigial teeth in badger (*Meles meles*), roe deer (*Capreolus capreolus*) and fox (*Vulpes vulpes*) from the county of Argyll, Scotland. *J. Zool., Lond.* **162,** 521–525.

READ, D. J. (1966). Dieback disease of pines with special reference to Corsican pine, *Pinus nigra* var. *calabrica* Schw. I. The nature of the disease

symptoms and their development in relation to the crown and to aspect. *Forestry* **39,** 151–161.

READ, D. J. (1967). *Brunchorstia die-back of Corsican pine.* Forest Rec., Lond. 61.

RISHBETH, J. (1950; 1951a and b). Observations on the biology of *Fomes annosus*, with particular reference to East Anglian pine plantations. I. The outbreak of disease and ecological status of the fungus. *Ann. Bot.* **14,** 365–383 (1950). II. Spore production, stump infection, and saprophytic activity in stumps. *Ann. Bot.* **15,** 1–21 (1951a). III. Natural and experimental infection of pines, and some factors affecting severity of the disease. *Ann. Bot.* **15,** 221–246 (1951b).

RISHBETH, J. (1952). Controls of *Fomes annosus* Fr. *Forestry* **25,** 41–50.

RISHBETH, J. (1957). Some further observations on *Fomes annosus* Fr. *Forestry* **30,** 69–89.

RISHBETH, J. (1959a and b; 1963). Stump protection against *Fomes annosus*. I. Treatment with creosote. *Ann. appl. Biol.* **47,** 519–528 (1959a). II. Treatment with substances other than creosote. *Ann. appl. Biol.* **47,** 529–541 (1959b). III. Inoculation with *Peniophora gigantea. Ann. appl. Biol.* **52,** 63–77 (1963).

RIVERS, C. F., and CROOKE, M. (1960). Virus control of the sawfly (*Neodiprion sertifer* Geoff.). *Proc. 5th Wld For. Congr. Seattle* **2** (3c), 951–952.

ROWE, J. J. (1966). Deer research in the Forestry Commission. *Deer News* **10,** 34–37.

ROWE, J. J. (1967). *The Grey squirrel and its control in Great Britain.* Res. Dev. Pap. For. Commn, Lond. 61.

ROWE, J. J. (1969). Research on fencing for forest protection. *J. chart. Ld Ag. Soc.* **69** (2), 56–57.

ROWE, J. J. (1971). Prevention of damage by birds and mammals in forest nurseries. *Q. Jl For.* **65,** 148–157.

ROWE, J. J., and JACKSON, D. A. (1971). An investigation of the problems involved in managing pheasants in extensive upland spruce forests. *Forestry and Game, Proceedings of Game Conservancy Symposium,* Oct. 1971, 12–19.

SHORTEN, M. (1957a). Damage caused by squirrels in Forestry Commission areas, 1954–1956. *Forestry* **30,** 151–172.

SHORTEN, M. (1957b). Squirrels in England, Wales and Scotland, 1955. *J. Anim. Ecol.* **26,** 287–294.

SHORTEN, M., and COURTIER, F. A. (1955). A population study of the Grey squirrel (*Sciurus carolinensis*) in May 1954. *Ann. appl. Biol.* **43,** 494–510.

SKILLING, D. D. (1969). Spore dispersal by *Scleroderris lagerbergii* under nursery and plantation conditions. *Pl. Dis. Reptr* **53,** 291–295.

STICKLAND, R. E. (1967). Insect suction trap for collecting segregated samples in a liquid. *J. agric. Engng Res.* **12** (4), 319–321.

STOAKLEY, J. T. (1967). *Oviposition period of the Douglas fir seed wasp Megastigmus spermotrophus.* Res. Dev. Pap. For. Commn, Lond. 43.

STOAKLEY, J. T. (1968). Control of the Pine weevil, *Hylobius abietis* L., and of *Hylastes* species. *Forestry* **41** (2), 182–188.

TAYLOR, J. C., LLOYD, H. G., and SHILLITO, J. F. (1968). Experiments with warfarin for Grey squirrel control. *Ann. appl. Biol.* **61**, 312–321.

TAYLOR, K. D. (1963). Some aspects of Grey squirrel control. *Ann. appl. Biol.* **51** (2), 334–338.

TAYLOR, K. D., SHORTEN, M., *et al.* (1971). Movements of the Grey squirrel as revealed by trapping .*J. appl. Ecol.* **8** (1), 123–146.

THOMAS, P. H., *et al.* (1962–1970). Firespread in forest and heathland fuels. *Rep. Forest Res., Lond.* 1961, 105–108; 1962, 116–119; 1963, 108–112; 1964, 121–123; 1965, 124–130; 1966, 109–110; 1967, 167; 1968, 164; 1969, 162; 1970, 190–194.

THOMPSON, H. V., and PEACE, T. R. (1962). The Grey squirrel problem. *Q. Jl For.* **56**, 33–41.

TOWNROW, J. A. (1954). The biology of *Cryptostroma corticale* and the Sooty bark disease of sycamore. *Rep. Forest Res., Lond.* 1953, 118–120.

VARTY, I. W. (1956). *Adelges insects of Silver firs.* Bull. For. Commn, Lond. 26.

WALLIS, G. W. (1960). Survey of *Fomes annosus* in East Anglian pine plantations. *Forestry* **33**, 203–214.

WALLIS, G. W. (1961). Infection of Scots pine roots by *Fomes annosus. Can. J. Bot.* **39**, 109–121.

WOOD, R. F., and HOLMES, G. D. (1957). Chemical weed control: Control of vegetation in fire traces. *Rep. Forest Res., Lond.* 41–42.

WOOD, R. F., and NIMMO, M. (1962). *Chalk downland afforestation.* Bull. For. Commn, Lond. 34.

YOUNG, C. W. T. (1969). *Larch canker and dieback.* Leafl. For. Commn 16. Revised.

YOUNG, C. W. T., and STROUTS, R. G. (1972). Dieback of Corsican pine. *Rep. Forest Res., Lond.* 94–95.

Chapter 26

MENSURATION AND FOREST MANAGEMENT

As mentioned in Chapter 11, the present Management Services Division stems from the post-war Mensuration Section of the Research Branch built up by F. C. Hummel. Forest Mensuration has two aspects; the study of growth and yield in which the research component is important, and the compilation and presentation of statistics derived from such studies which represents a direct service to forest management. In the post-war period, with rapidly increasing production and the growth of new industries, emphasis shifted to the broader managerial problems of forecasting and control of yield for which mensurational studies had provided a firm basis.

The Mensuration Section, enlarged by the addition of Working Plans and Economics, and retitled "Management Services," left the Research Branch in 1960 to become a Headquarters Section. Though now designed to provide an expert service to management, there was still a considerable element of investigation in its work. Some of this was methodological, such as finding the best techniques for British conditions in conducting enumerations at the forest and regional level. Much work was also done with Utilisation Development on supply questions for new industries. However, the research side does shade into normal managerial investigations, and a full treatment of these would be out of place here.

Forest Mensuration

F. C. Hummel's stock-in-trade on his appointment as Mensuration Officer in 1946 consisted of the yield tables and the well-established procedure for conducting sample plots published in 1928 in James Macdonald's *Growth and yield of conifers in Great Britain* (Macdonald, 1928). The yield tables in this work were compiled mainly from temporary plots measured towards the end of the first war; data from the Commission's own permanent plots being used in revision and for the provisional tables for Sitka spruce. Some 270 permanent plots in all species were in existence in 1946. The position was reviewed and new permanent plots were added to improve the coverage of species and growth classes at an average rate of some fifty plots a year over the next decade. The procedure for conducting sample plot work was revised over the years and eventually published as Forestry Commission Bulletin No. 31 *Code of sample plot procedure* (Hummel *et al.*, 1959).

An early objective was the revision of the pre-war *Yield tables* and preparation of new tables for the less commonly-used species; provisional tables for Japanese larch, which was now being planted on an increasing scale, were published in 1949 (Hummel, 1949). However, the immediate post-war priority for Mensuration was in connection with the Census of Woodlands (Forestry Commission, 1952–53). The Census, which had been initiated by O. J. Sangar before the war, was restarted and completed by J. S. R. Chard in the period 1947–1949. It was a comprehensive survey and classification of all woods of five acres or more in extent. To obtain regional and national statistics of volume and increment, sampling procedures were worked out with the advice of F. Yates of Rothamsted Experimental Station, and P. J. Finney of the Imperial (Commonwealth) Forestry Institute (Finney, 1948; Finney and Palca, 1949). Provisional volume tables were prepared to assist in computing stand volumes, and these formed the basis of the later published versions. Increment was estimated from available yield tables, or by direct methods; for broadleaved trees for instance it was necessary to resort to stem analyses.

The Census of 1947–49 was followed by a survey of small woods, parkland and hedgerow timber (Census Report No. 2, 1953). It was originally intended to revise the Census by re-survey on the same lines, county by county, but this was considered unprofitable, and the next national survey (which was restricted to private woodlands) conducted in the mid-sixties was based on random sampling at 15 per cent level, the kilometre grid square being the unit. Inside this unit, sampling for volume etc. was stratified by crop types. This survey was of special interest as it was the first major experience in British forestry in capturing numerical data for automatic processing. Special forms were used for processing by the English Electric reading device "Lector," which printed the data to punched tape for the computer. Results were published by HMSO in *Census of woodlands 1965–67* (Locke, 1970).

The preparation of volume tables was based on the well-known near-linear relationship between stem volume and breast-height basal area, from which tables have often been prepared simply from graphs. Hummel made increasing use of regressions, and at this time some compromise was necessary in the number of terms in the interests of ease in calculation. This is of course a type of work in which the computer is a great relief. During his work on volume tables, Hummel was impressed by the regularity in which volume/basal area lines intercepted the x axis at a constant value for basal area, and this gave him the idea of constructing local volume tables from tribes of lines passing through this point, and separated by unit intervals of volume and basal area. In using such tables (known as tariff tables), the appropriate line for a stand is found by sampling a relatively few trees, the volumes of any others in the stand being read off from the appropriate table. Hummel's work in this particular field of mensuration is described in his *Volume-basal area line*, Forestry Commission Bulletin No. 24 (Hummel, 1955). Tariff tables, which are accurate enough for most purposes, have been in general use since the mid-fifties and have proved a great convenience. General volume tables for individual species were prepared by Hummel with Irvine, Christie, Jeffers, Waters, and others; appearing in the Commission's series of Forest Records through the fifties. (*Detailed in Check list of Forestry Commission publications, 1919–65*, For. Rec., Lond. 58, HMSO.)

The first revision of Macdonald's pre-war *Yield tables*, by Hummel and Christie, was published in 1953 (Hummel and Christie, 1953). These tables were still prepared by traditional empirical methods based on hand-fitted curves; the main changes lay in the earlier, and rather heavier, thinning regimes, and the Quality Class of stand was now based on the top height (mean of 100 largest trees per acre) rather than the average. These were followed in the period 1957–61 by provisional Yield Tables for some minor coniferous species, Western hemlock, Grand fir, Noble fir, as well as tables for oak, beech and poplar. The most recent Yield Tables and other complementary tables (Management Tables) have however followed definite changes in ideas about thinning and classification of rates of growth of the stand.

Thinning practice in the post-war period attracted much argument, with strong advocacy for particular types. It is not possible to review the subject here and only the experimental evidence will be mentioned. The standard thinning grades used in sample plots before the war, and for some time after it, were controlled by the dominance class of the trees amongst which thinning took place, and secondly by the weight of thinning; sample plots thinned to different regimes thus presented contrasts in stand structure and intensity of thinning. Replicated thinning experiments were rare, in fact the Bowmont experiment established in 1930 remained unique till the early sixties. However, evidence on thinning method and yield was obtainable from the records of numerous plots thinned on different regimes, and the manner in which they fitted into the general curves for increment over age suggested that (within the limits examined) there was little effect of thinning type or intensity on production. In fact, when the precise Bowmont experiment showed significant (though not very large) differences in favour of the two heaviest intensities, it was a matter of some surprise. Perhaps one of the most important points about the Bowmont experiment was that the heavy,

low, thinning compared favourably with the light crown thinning in volume yield. The latter type of thinning was designed to produce deep, storeyed canopy, thought especially suited to a tolerant species such as Norway spruce, and though this effect can be accentuated by heavier crown thinnings, the indications at Bowmont did not encourage the idea that production would be increased by departing from uniformity of stand structure. And bearing on this point, A. M. Mackenzie (1962) made the interesting observation that the crown surface area in all the thinning treatments at Bowmont was much the same, about three times the area of the ground occupied by the crops.

Guillebaud and Hummel (1949) looked at the records of individual trees of various dominance classes in permanent sample plots in pure even-aged stands, and found as they expected that the general trend was downwards. As this implies that a crop in later life consists almost entirely of the original dominants, it points fairly obviously to rigorous early selection amongst the dominants in early thinnings, and though it has no bearing on intensity or kind of thinning, it does not encourage any fussy treatment of the lower dominance classes.

From 1956 Mensuration formed part of an enlarged Section concerned with the wider subject of Forest Management, and in 1960 this Section became attached to Headquarters. The presentation of information derived from mensurational studies reflects the change. For example, the *Forest management tables* prepared by R. T. Bradley, J. M. Christie and D. R. Johnston (1966) and published as Forestry Commission Booklet No. 16 (since revised in metric units as F.C. Booklet 34, HMSO) was designed specifically to aid yield control and production forecasting as well as to the more traditional requirements which yield tables provided. This publication broke new ground in several ways. The growth of stands was now classified primarily by the maximum mean annual increment, termed Yield Classes, as against the old Quality Classes which represented intervals in top height at a given age, and did not allow comparisons of yield between species. Height, however, remained the most practical indication of the potential maximum mean annual increment. The concept of Production Class was introduced to accommodate local variations in the relationship between height growth and volume production.

The procedures advocated in yield control contained a change in emphasis from controlling the growing stock after thinning to controlling the volume of thinnings removed. The thinning policy embraced a thinning intensity approximating to the marginal thinning intensity, i.e. the maximum thinning intensity which could be applied without risking loss of production. Furthermore, the thinning regime adopted provided for a constant annual thinning yield over the thinning life of the crop which was equivalent to 70 per cent of the maximum mean annual increment. This prescription had been arrived at from a study of the records of numerous permanent sample plots including thinning experiments.The attainment of a common quantitative standard has undoubtedly been a great assistance to management in the forecasting and control of production, in contrast to the earlier concept of thinning grades which was found to be incapable of consistent application (Bradley, 1967a).

Further replicated thinning experiments in several species have been established since the sixties. Of principal interest has been the study of the effect of intensity of thinning on production, embracing levels of intensity beyond the marginal thinning intensities. In addition, the secondary aspects of thinning type (crown or low etc.), thinning cycle, and thinning patterns such as line thinning, have been incorporated in the treatments comprising these experiments. In the early sixties, control of these experiments was by the proportion of basal area removed, though later control has been in terms of thinning intensity as described above (Bradley, 1967b). (Plates 68, 69.)

Efforts have been made to speed up the collection of information by more intensive assessment: for example, an experiment in thinning of Douglas fir at Alice Holt used vernier girth bands to measure the response to thinning at short intervals through the season. Similar techniques were applied to the study of individual trees subject to release by the removal of competitors at the Forest of Ae. An interesting idea derived from Japan is the use of nursery stock as a model forest in thinning experiments. This has not been taken far since, although a few small experiments indicated certain similarities between transplants and older trees regarding the effects of thinning, the fundamental differences, notably in crown structure, between the physical features of transplants and older trees tends to limit the application of the results of this type of experimentation.

Most sample plots and thinning experiments have been laid down in normally spaced crops, and have given no information on the effects of initial spacing on production. Plots, however, were established in James Macdonald's 1935 spacing experiments which range from 3 ft × 3 ft (0.9m) to 8 ft × 8 ft (2.4m) in initial plant spacing. Generally speaking the evidence is that by about the normal time of first thinning there is a slight "initial" loss of production of 15–40m³/ha in the widest spacings relative to the closest, after which the yield follows the usual pattern, being much the same in fully stocked crops whatever the initial spacing.

As experimentation in spacing and thinning covers more extreme treatments, the question of timber quality becomes more important, and this is an area where little hard fact is available.

Towards the end of the sixties a major advance in yield table construction was made possible by the mathematical characterisation of most of the basic growth function used in producing the Forest Management Tables (Christie, 1972; Hamilton and Christie, 1971). The end-product of this development is the facility to produce by computer yield models of a given species and yield class subjected to a specified thinning treatment.

In addition to work on mensurational methods, attention was paid to instrumentation. Probably the most important item was the development of an optical dendrometer for accurate diameter measurements of standing trees, based on the principles of military range finders. This instrument, described by J. N. R. Jeffers (1956), was finally developed and marketed by Messrs Barr and Stroud. Bitterlich's ingenious angle gauge or relascope was introduced about 1950, and various applications worked out for it (Finch, 1957; Keen, 1950). It gives an estimate of basal area from a count of the number of stems, in a 360° sweep, subtending angles over certain values at the instrument. (Plates 70 and 71.)

There has always been argument about the relative advantages of tapes versus calipers, i.e. measurement of girth versus diameter as a stage to basal area. On the whole the British preference for tapes has been maintained. An interesting development of the caliper in the early sixties, due to René Badan, a visiting Swiss graduate, incorporated a portable tape puncher with the caliper to record diameter measurements directly in a form suited to computer input (Badan *et al.*, 1961). This has now been developed further and is being marketed commercially under the trade name of 'la bastringue'.

Working Plans

D. R. Johnston was appointed in 1956 to this new Section, which employed specialist teams of Forester surveyors, trained with assistance from the Ordnance Survey, to collect the standard data required for planning at all levels of management. The Commission's decision to introduce formal working plans raised a number of questions of a managerial nature which need not be discussed here. Technical problems were concerned with such matters as enumeration and production forecasts, in which the Mensuration Section had already gained experience in the previous decade. Map construction and reproduction received considerable attention and the use of available air photographs, mainly Royal Air Force source of mixed quality, indicated their potential use for stock mapping and other applications. The cost, however, of direct contract flying to obtain specific air photo coverage outweighed any benefits to the Commission. The question of soil mapping in forestry was raised at an early stage. The national Soil Surveys of England and Wales and Scotland, based on Rothamsted and the Macaulay Institute respectively, and carried out under the direction of the Agricultural Research Council, were by no means completed at this time, and had naturally concentrated on agricultural soils. A great area of potential forest land had not been surveyed in any detail. In beginning its own soil survey work in 1960, the Commission was indebted to the Soil Survey of England and Wales for training of staff, and based its methods of classification on those of the Survey, i.e. the stratification and physical factors of the profile. D. G. Pyatt's surveys were concerned with two main topics. In certain forests where plans were under consideration, the original choice of species had not made the best use of soil types, and here soil survey contained an element of enquiry into the site requirements of species and their site/growth relationship in order to determine those factors, possibly other than the soil series classification, which may be the best criteria for site mapping. Secondly, and of more importance, there was the question of stability of crops against windthrow in relation to soil conditions. On this topic Pyatt worked closely with the Research Sections (Soils and Silviculture) more concerned with this aspect of site (as mentioned in Chapter 21), and the topographic/edaphic classification of forest areas according to the hazard of windthrow is of obvious value in considering the rotation length in vulnerable forests.

Forest Economics
The application of economic theory to forestry is by no means new, and may be dated from Faustmann's treatment of capital, rental and soil expectation value in the nineteenth century. In Britain, W. E. Hiley had been the most prominent exponent of ideas on economic management of forests till his death in 1961, and the subject was necessarily represented in the University faculties of forestry. The appointment of an economist (A. J. Grayson) in 1956 was an essential addition to a team which was increasingly concerned with management policy (Johnston *et al.*, 1967). Grayson and his colleagues have been engaged in a wide range of enquiries which can only be briefly scanned here (Grayson, 1961; 1967a; 1968). Some of the work has been concerned with the provision of economic statistics such as prices and consumption of wood and wood products (Grayson, 1966; 1969), and the development of costs and labour productivity. Such statistics and related forecasts have provided material for bodies concerned with forest policy or land use at the national level, such as interdepartmental working parties on Forestry Policy and the Land Use Study Group.

A type of enquiry which has been pursued over a wide range of forest activities compares the profitability of different courses of action, or different intensities of effort. The net discounted revenue (the difference between prospective future returns and costs all discounted to the present time) either per unit area or per £ invested, has been adopted as the standard criterion in much of this work. Topics studied have included road densities, choice of species, spacing, thinning regime, felling age, fertilisation and many others (Grayson, 1967b). This approach has emphasised the economic importance of relating the costs of increasing production to the revenue implications of such production.

Applications of the criterion of maximum net discounted revenue have led to more critical appraisal of forest management options in the field, and have also influenced research thinking. Yield forecasting since 1965 has been carried out by computer, thus enabling speedy calculation of the result of applying a variety of cutting regimes to data on growing stock compiled by the Field Surveys Branch (formerly Working Plans Branch).

Armed with such tools, Economics moved, from the mid-sixties, to broader enquiries concerned with planning at the regional level. The principal reasons for these studies in applied economics were, firstly, to ensure that consistent decisions were being formulated in all operations and management units, and secondly to ensure that interactions between different units were explicitly taken into account in planning (Wardle, 1965). Some of these enquiries have dealt with particular projects such as the supply of wood from a number of forests to a particular pulp mill and the planning of sowing, lining-out and plant allocation (Grevatt and Wardle, 1967; Wardle, 1971). Others have dealt with the management of a single unit of the Commission's enterprise, namely a Conservancy.

Considering the management of the Commission's forest enterprise as a whole, a useful tool has been developed in the form of a computer simulation programme. This identifies the major results, in terms of output and resource requirement, of alternative programmes of work and intensities of operation. Such explorations have proved invaluable in certain areas of corporate planning, a field with which economics has been increasingly concerned.

The sixties saw a major growth of interest in questions of recreation and amenity. Economics has been concerned in particular with assessing the numbers of visitors to forests, and the valuation of recreation in money terms (Sidaway, 1970).

Other developments in which Economics (or, as it has been known since 1969, Planning and Economics) has played a part are the establishment of a system of internal financial control for the Commission and the formulation of a new method of financing and reviewing the financial achievements of the Commission.

REFERENCES

BADAN, R., HINSON, W. H., and STEWART, D. (1961). Un compas forestier enregistreur: nouvelle contribution à l'application de la mécanographie dans les inventaires forestiers. *Proc. 13th Congr. int. Un. Forest Res. Org.*, Vienna, 2 (25/7). 2 pp.

BRADLEY, R. T. (1967a). *Thinning control in British woodlands*. Bookl. For. Commn 17.

BRADLEY, R. T. (1967b). *Thinning experiments and the application of research findings in Britain*. Res. Dev. Pap. For. Commn, Lond. 63.

BRADLEY, R. T., CHRISTIE, J. M., and JOHNSTON, D. R. (1966). *Forest management tables*. Bookl. For. Commn 16. See also Hamilton and Christie, below.

CHRISTIE, J. M. (1959). *Preliminary yield table for poplar*. Forest Rec., Lond. 40.

CHRISTIE, J. M. (1972). The characterization of the relationships between basic crop parameters in yield table construction. In *IUFRO 3rd Conference Advisory Group of Forest Statisticians, Juoy-en-Josas, 7–11 September 1970*, pp. 37–54. Paris: Institut National de la Recherche Agronomique.

CHRISTIE, J. M., and LEWIS, R. E. A. (1961). *Provisional yield tables for Abies grandis and Abies nobilis (A. procera)*. Forest Rec., Lond. 47.

EVANS, W. R., and CHRISTIE, J. M. (1957). *Provisional yield table for Western hemlock in Great Britain*. Forest Rec., Lond. 33.

FINCH, H. D. S. (1957). Plotless enumeration with angle gauges. *Forestry* 30 (2), 173–192.

FINNEY, D. J. (1948). Random and systematic sampling in timber surveys. *Forestry* 22 (1), 64–99.

FINNEY, D. J., and PALCA, H. (1949). The elimination of bias due to edge effects in forest sampling. *Forestry* 23 (1), 31–47.

FORESTRY COMMISSION (1952–53). *Census of woodlands 1947–49*. In 5 parts. London: HMSO.

GRAYSON, A. J. (1961). Influence of log size on value: A report on a study of the Cowal-Ari mill, Strachur, Argyll. *Timb. Trades J.*, 2 Sept., 75–78.

GRAYSON, A. J. (1966). Economic statistics required for policy formulation. *Proc. 6th Wld For. Congr.* (3), 3885–3890.

GRAYSON, A. J. (1967a). *Afforestation planning at the national and project levels*. Res. Dev. Pap. For. Commn, Lond. 35.

GRAYSON, A. J. (1967b). Species, growth rate and profitability. *Timb. Grow.*, Jan., 20–27.

GRAYSON, A. J. (1968). *The formulation of production goals in forestry*. Res. Dev. Pap. For. Commn, Lond. 56.

GRAYSON, A. J. (1969). *Imports and consumption of wood products in the United Kingdom 1950–1967, with forecasts to 1980*. Forest Rec., Lond. 70.

GREVATT, J. G., and WARDLE, P. A. (1967). *Two mathematical models to aid in nursery planning*. Res. Dev. Pap. For. Commn, Lond. 44.

GUILLEBAUD, W. H., and HUMMEL, F. C. (1949). A note on the movement of tree classes. *Forestry* 23 (1), 1–14.

HAMILTON, G. J., and CHRISTIE, J. M. (1971). *Forest management tables (metric)*. Bookl. For. Commn 34.

HUMMEL, F. C. (1949). *Revised yield tables for Japanese larch in Great Britain*. Forest Rec., Lond. 1.

HUMMEL, F. C. (1955). *The volume-basal area line*. Bull. For. Commn, Lond. 24.

HUMMEL, F. C. (1956). *Tariff tables for conifers in Great Britain*. Forest Rec., Lond. 31.

HUMMEL, F. C., and CHRISTIE, J. (1953). *Revised yield tables for conifers in Great Britain*. Forest Rec., Lond. 24.

HUMMEL, F. C., LOCKE, G. M. L., JEFFERS, J. N. R., and CHRISTIE, J. M. (1959). *Code of sample plot procedure*. Bull. For. Commn, Lond. 31.

HUMMEL, F. C., LOCKE, G. M. L., and VEREL, J. P. (1962). *Tariff tables*. Forest Rec., Lond. 31. Revised.

HUMMEL, F. C., et al. (1951). *General volume tables for oak, beech, birch, Scots pine, European larch, Norway spruce, Corsican pine, Japanese larch and Douglas fir in Great Britain*. Forest Rec., Lond. 5–11 and 14–15 respectively.

JEFFERS, J. N. R. (1956). Barr and Stroud dendrometer, Type FP7. *Rep. Forest Res., Lond.* 1955, 127–136.

JOHNSTON, D. R., GRAYSON, A J., and BRADLEY, R. T. (1967). *Forest planning*. London: Faber & Faber.

KEEN, E. A. (1950). The relascope. *Emp. For. Rev.* 29 (3), 252–264.

LOCKE, G. M. L. (1970). *Census of woodlands 1965–1967*. Forestry Commission. London: HMSO.

MACDONALD, JAMES (1928). *Growth and yield of conifers in Great Britain*. Bull. For. Commn, Lond. 10.

MACKENZIE, A. M. (1962). The Bowmont Norway spruce sample plots 1930–1960. *Forestry* 35 (2), 129–138.

SIDAWAY, R. M. (1970). Measuring forest recreation. *Suppl. Timb. Trades J.* Oct., 37–39.

WARDLE, P. A. (1965). Forest management and operational research: A linear programming study. *Mgmt Sci.* **11** (10), B260–B270.

WARDLE, P. A. (Editor) (1971). *Operational research and the managerial economics of forestry.* Bull. For. Commn, Lond. 44.

WATERS, W. T., and CHRISTIE, J. M. (1958). *Provisional yield tables for oak and beech in Great Britain.* Forest Rec., Lond. 36.

Chapter 27

OPERATIONAL RESEARCH AND DEVELOPMENT

By no means all the activities which can be classed under this head have been so labelled and organised; a great deal of the work has simply fallen to staff in the field or at headquarters in line with their duties. It is difficult to draw hard and fast lines, but preference has been given here to work in the Research organisation, or which has been linked with it in some obvious way.

Mechanical Development

Established in 1949 under Col. R. G. Shaw, this was the first of the Sections concerned primarily with development work. Col. Shaw answered to the Director of Research, but the work was supervised by a Headquarters Committee (the Mechanical Development Committee), which had the task of coordinating development projects outside Research as well. Mechanical engineers stationed at the old territorial Directorate headquarters were also concerned, and many field officers from foresters upwards took part from time to time in development work.

The Work Study Section became concerned in testing of machinery on operations in collaboration with Col. Shaw in the early sixties, and after Col. Shaw's retirement in 1965, Work Study (by now incorporated in the Management Services Division) took over mechanical development altogether.

One or two generalisations may be made about the Commission's work in mechanical development. Relatively little of the equipment which has been brought into use has been designed initially for home forestry. Much of the work has been adaptation or modification of existing equipment to suit British conditions, and in some important fields (such as power saws and machines used in the extraction of produce) development has been on the international level. However, there are several instances of design and development purely for British conditions, the best examples being in ploughing equipment.

The main developments in ploughing machinery have been mentioned in Chapter 9, since they are very closely associated with experimental work in the establishment of plantations. Tractors have of course greatly increased in power from the standard, wheeled, agricultural models first used. American crawler types were first used in ploughing just before the war, and in the immediate post-war years efforts were made to find suitable British-made crawlers to replace the American types. At the same time a good deal of attention was given to adaptations for the softest ground conditions, such as low track-loading, forward centre-of-gravity etc. The Fordson "Long-Wide County" was developed by Commission

engineers and the manufacturers, and remained the standard soft ground tractor from the mid-fifties till it went out of production in 1967. The increasing capacity of ploughing equipment required much work on mounting and linkage systems, and both carriages and direct linkage systems are now in use.

For the very deepest drains of 30 inch or more it is usually necessary to resort to back-acting diggers, which are expensive to operate. It had been hoped that the Finnish Lokomo plough, which is winch-pulled from an anchored tractor, would be able to drain to this sort of depth, but it has limited scope. An operation which has proved extremely difficult to mechanise is that of cleaning drains, especially those in plantations. Various approaches have been made. Shaw constructed a machine operating as a flail, and a German device was tried which worked on the screw principle, but on the whole there has been little success, and the hope is that modern well-aligned main drains will be self-cleaning.

The other main field of mechanical development has been in everything concerned with harvesting and extraction. Little need be said about the power saw, a truly international development, but near the end of the period the Commission began to play a part in developments towards decreasing noise and vibration in saws, both of which had been shown to have harmful effects on men using saws continuously. As technology has succeeded in doubling the speed of cut and halving the weight of saws inside a decade, it is to be hoped that power saws can also be made less objectionable.

There is nothing very special about the problems of extraction in British forestry, though the handling of small produce has been of more significance here than in most other countries. The great effort in road construction, well described by Ryle (1969), which was started in 1945, has of course been the greatest single factor. In the early post-war period there was some interest in systems such as portable chutes and cableways for sparsely roaded hill forests, but all trends have been away from very long off-road hauls. In the primary extraction of small thinnings it has taken some time to replace the horse, but the arrival of winch systems such as the Isaachsen have provided more power and nearly as much flexibility. A great deal has been done in the testing and development of tractors for off-road haulage of heavier thinnings. Starting with the more-or-less standard agricultural model fitted with a single drum winch, the forest tractor has evolved rapidly in this period. Improvements in traction from the introduction of four-wheeled drive and large diameter tyres have enabled the wheeled tractor to compete successfully with the more expensive crawler in many situations. Frame steering, i.e. steering by turning the frame of the machine about a central pivot by hydraulic ram, is a mechanical principle which appeared in foreign machines in the early sixties. Perhaps the most exciting innovation is hydrostatic transmission, first developed by the National Institute of Agricultural Engineering at Silsoe, Bedfordshire, which dispenses with clutch and brakes and gives continuously variable drive. It has not yet reached full production standards.

Loading of logs and poles, whether to tractor/trailer combinations (forwarders) "off-road" or to road vehicles, has been greatly facilitated by the mounted hydraulically operated hoists, such as the Swedish HIAB, which arrived in the early fifties, and by the Volvo front-end loaders also from Sweden.

The mechanisation of debarking of roundwood presented some problems. Small machines appeared which were suitable for working close to thinning operations, but the trend has been to debark at larger depots using more powerful machines, such as the Swedish Cambio. It is now common for pulp manufacturers to do their debarking at the plant. One obdurate problem has been the debarking of crooked poles for pulpwood.

Other forest operations which received attention included the clearance of scrub. Various machines which were tried out by A. D. S. Miller in the Derelict Woodlands project have already been mentioned in Chapter 19. Several small powered machines derived from mechanical hedge cutters have been tested for plantation weeding, but the most interesting developments follow the principles of the rotary forage cutters. Woodland versions of this type of machine are capable of comminuting low scrub or felling slash. They have been adapted for weeding in plantations also, and providing the plantation has been laid out for mechanical weeding they are cheaper than herbicides, and much cheaper than hand weeding. Mechanical Development has also been concerned with the design of spraying equipment of various kinds.

In the nursery, the development of machines has been a curiously unco-ordinated process, reflecting the diversity of nursery conditions. The basic cultural operations up to the preparation, sowing (though still often done by hand) and grit covering of seedbeds, have become standardised about the Ferguson tractor and a range of appliances mounted to the tool bar. Whilst broadcast sowing holds the field at present, drilling (long mechanised in agriculture) has been an obvious target. Machines for drilling or band sowing conifers have been constructed (notably by M. E. McNulty (1951) and A. Slatter). There are undoubtedly technical diffi-

culties connected with dispersing the sowing of small-seeded, relatively slow-germinating species, over greater nursery areas—as against concentrating in the specially prepared bed. These could probably be mastered, and there would be a stimulus to do so if the use of undercut stock became general, for there are obvious attractions in running a nursery on a row-crop system throughout. Undercutting, however, though it has been shown to produce a satisfactory plant, has been applied mainly to beds, and though this can be done, it is not mechanically as easy as undercutting rows. Transplanting is still fairly general, and one valuable approach to mechanising the operation has been the lining-out plough, developed by A. Rose (1963) at Ledmore Nursery, which combines the work of taking out the furrow for the lining-out boards with filling and firming the soil against the previous "line".

Transplanting machines, derived from types designed to plant cabbage and other vegetable seedlings, have had extensive trials. It is rather difficult to attain the normal close spacing by machine, and the planting action is apt to result in lopsided roots. No doubt on perfect soils this process could be satisfactorily mechanised, but in most of our nurseries it seems likely that we shall stop at something less sophisticated. In the early post-war period efforts were made (with some success) to adapt agricultural row-crop weeding devices for use in transplant lines. Mechanical weeding has, however, been by-passed by the modern herbicides.

Work Study

This form of operational research originated in American industry, but the Swedes have led in applying it to forestry. The Commission's Work Study Section was established in 1954 under J. W. L Zehetmayr, who was succeeded in 1964 by L. C. Troup. It was loosely attached to Headquarters, with no formal links with Research. The approach and techniques being new to forestry in Britain, special training for the staff was necessary. The Section has worked from the outset in small teams of a leader and two to five others.

The early Work Study projects were mainly in production, which was increasing rapidly during this period, though high costs gave rise to some anxiety. Projects centred on a particular category of produce, most of which were converted from softwood thinnings. Besides the traditional outlets such as the various classes of pitwood, markets were appearing for pulpwood and chipwood, and thinnings in the older plantations were beginning to yield a proportion of sawlogs. In the early stages it was necessary to pay close attention to tools, methods of using them, and maintenance. Experience gained in this was built into the training system.

Work study in forest operations (or any other) requires a detailed investigation of the processes, the timing of various stages, and measurement of the quantities involved; the object being to optimise the use of time, effort and materials. Through the fixing of standard times for various parts of an operation it has been used to calculate more logical and fairer piece-rates, and this alone has had effects on productivity. It has also provided the basis for more realistic costing and production control, and has yielded a great deal of information on factors governing the economy of working. For instance, in thinning operations frequent light thinnings have been shown to be very uneconomical, and there is also a steep relationship between unit costs and size of tree. This has influenced weight and periodicity of thinning in practice. It has also provided a good basis for comparing the profitability of different specifications, where there is a choice, and one of the guiding principles which has emerged is "minimal conversion at stump".

The main development of the sixties has been the broadening of the approach to production operations to include the whole range of extraction problems. Whereas, to start with, Work Study collaborated with the Machinery Research Officer in the field trials of various new appliances, after Col. Shaw's retirement a mechanical engineer (R. B. Ross) joined the Work Study team and by 1967 mechanical development had passed to Work Study. The linkage with Planning and Economics in the Management Services Division also had its effect on widening the scope of the Branch. In addition to assignments on particular types of produce (some of which, such as the supply for the Scottish Pulp paper mill at Fort William, have been taken "from the stump to the chipper"), whole systems and the machinery required for them have been studied. The two main systems—shortwood and whole tree—require a different spectrum of machines, and the work of the last few years has gone a good way towards determining the best combinations. It is of interest that Work Study, which started life as a non-experimental Section, has since found it profitable to add experimental teams to its strength, especially for work on new machines and techniques.

Production and extraction have been the main fields for Work Study, but in the sixties it had become increasingly concerned with silvicultural operations, especially those with a high labour content. While there is often some difficulty over the job specification in silvicultural work, there are many cases where treatments and their effects are known in quantitative terms, but the means of carrying them out have not been so well studied, and here the Work Study approach can add a new dimension. For instance even in a practice as old as planting,

where the conditions required are well known, there have been continuing differences in opinion about tools, methods, and organisation of work. This is an obvious enough topic, and Work Study can contribute also on the development of a new technique, such as the use of tubed seedling stock. Weeding provides an interesting case, since Silviculture has had considerable successes in developing the use of modern herbicides, but not so much effort has been put into machines. Work Study has designed machines based on the most promising principles of those already available, and recent trials suggest that mechanical weeding can compete with chemical weeding under some conditions, though of course the ultimate test is not short-term, since machines and herbicides are doing different things.

The study of an actual operation gets more complicated when it is part of a whole system. In thinning, for instance, there has been renewed interest in the old idea of line thinning, which was practised during the war in East Anglian pine forests, and has obvious advantages for extraction. This however is not a matter of operational efficiency alone, since the health, stability, yield and quality of crops also have to be considered, and some way has to be found to evaluate these criteria short of embarking on vast factorial experiments.

Current programmes give the impression that Work Study is growing into a development Section complementary to Research. There is no doubt there is a need for such activity, and whether it is better associated with management rather than research is a matter of convenience. Much of our experimental work however being purely practical in its objects, it might well be useful to bring in operational research techniques at the outset, using mixed project teams.

Utilisation Development

A small section under E. G. Richards was established in 1950 to carry out investigations into utilisation, the Commission being advised by a Committee on the Utilisation of Home Grown Timber, which had been brought together the previous year. The statutory Advisory Committee on Home Grown Timber later took over its functions. Richards was succeeded by B. W. Holtam in 1966, and when Holtam was promoted to the new post of Chief Research Officer (North) at Edinburgh in 1968, J. R. Aaron took over from him. Since 1965 Utilisation Development has been a Section of the H.Q. Harvesting and Marketing Division.

The work of the Section has been a mixture of development and research, much of the latter being initiated by the Section but carried out by the Forest Products Research Laboratory, Princes Risborough (now the Princes Risborough Laboratory of the Building Research Establishment, Department of the

Environment) and by other specialised laboratories connected with the trade.

In the early fifties priority was given to finding markets for small conifer thinnings and small roundwood from broadleaved coppice and scrub. In this, and subsequent, marketing enquiries, the primary task has been to get estimates of supply through the information provided by the Census of Woodlands (completed in 1949) and the more detailed statistics available for the Commission's own production. Several special enquiries were mounted in connection with supply to the growing pulp industry. The hardwood pulp plant established by Wiggins Teape Ltd. at Sudbrook, Monmouthshire, in the mid-fifties, required a great deal of preliminary investigation owing to the variety of species and types of roundwood (coppice, scrub, branchwood, etc.) available. Extensive surveys were conducted to obtain reliable values for the specific gravity and moisture content of the common conifers for the estimation of pulp yield and the pricing of supplies. The variation inside a given species in these characters was found to be very considerable.

Market surveys into consumer industries for both round and sawn softwood were undertaken in the fifties. These included such outlets as packaging and materials handling, and house construction; they sought information on the dimensions, grades and standards of preparation required for home-grown softwoods to compete in these markets. Other marketing enquiries were conducted on minor products, such as sawmill wastes, bark and extractives (tannin etc.) (Aaron, 1970). The use of some potentially valuable waste substances such as sawdust is handicapped by the relatively small size and scatter of the mills, but the modern pulp plants turn out large quantities of bark. As a result of recent development work substantial quantities of pulverised coniferous bark are now marketed for various uses in horticulture.

As previously mentioned, the Forest Products Research Laboratory has always included home-grown timber in its programmes, but in the post-war period it was able to devote relatively more time to it. One of the more important assignments of the late fifties was an investigation into the strength of pit-props from home-grown conifers, which was carried out in association with the Commission and the National Coal Board. It was found that home props prepared to specification and properly seasoned (the importance of this was emphasised in the enquiry) were not inferior to imported material. During the fifties the opportunity was taken to look at the properties of some of the exotics whenever mature material became available. Whilst this gave useful preliminary information on general characteristics, there was the disadvantage that pioneer plantings were apt to be untypical; the sites for instance were usually too good, and the trees were sometimes specimens rather than forest grown. By the late fifties, however, the earliest Commission plantations were beginning to produce saw timber, and it was decided to launch a major programme of investigation into the properties of home-grown timbers. The programme (which continues) started in 1960, and has been a joint effort between the Laboratory and the Commission, directed by a working committee representing both organisations. Compared with previous work on home-grown timbers at the Laboratory, it was a departure in that many of the questions were posed from the forester's standpoint; e.g. to find what were the effects on timber properties of differences in site, silvicultural practice, and inheritance (Harding, 1969–72).

It was also thought important to know about the variations in timber properties at the between-tree and within-tree levels, but the relatively large samples required for testing some of the gross characters (milling, planing, grading out-turn etc.) led to a division of the programme between extensive tests of timber properties and more intensive work on wood anatomy, in which it was possible to study the whole pattern of variation, including that across the growth ring. Sitka spruce was naturally given the first priority in the extensive studies, and samples were drawn from 30 to 37-year-old plantations selected on a regional basis. Plantation-grown material of this age-class showed a number of features connected with juvenility and the wide inner rings; such as low density (especially about the fifteenth ring) spirality on grain with twist of drying, and tendency for the widest ring material to "washboard" on kiln seasoning. However, these disadvantages are mainly associated with the juvenile core, there being a progressive improvement in density and fibre angle outwards, and they have not precluded young Sitka spruce from the market for saw-logs. It is of course greatly in its favour that it is a first class pulping timber, and much of the younger material is now pulped.

A number of the other exotics were subjected to standard tests. Age for age, Western hemlock and *Abies grandis* compared quite favourably with Sitka spruce, and Lodgepole pine showed good properties. In some species, comparative studies with competitors appeared the most informative line. Japanese larch was compared with the much more familiar timber of European larch, and age for age appeared to have much the same properties. Corsican pine likewise showed no inferiority to Scots pine in timber properties, with a better log form and higher out-turn from the same sites.

No great amount of material from thinning experiments has yet been put through, but timber became available from a Scots pine trial 74 years old subjected to different thinning grades for about forty years, and was converted to measure the out-turn of sawn material falling in the various grades under the Laboratory's standard rules. At this age, the advantage was clearly with the heavier thinnings, since any benefit due to narrower ring or smaller knots in the lighter thinnings was cancelled by the low out-turn of sawn material.

More, however, remains to be done on the effects of thinning and the related subject of initial spacing. Spacing experiments in Sitka spruce have been used to study density and anatomical characteristics associated with contrasting radial increment on the same site, and there seems no doubt there is some adverse effect on basic properties due to the increased proportion of springwood in the wide ring. The difficulty is to put a value on such effects; to balance them against the undoubted advantages of reaching given diameters of log quickly.

The effects of pruning have also been studied by standard conversion and measurement of out-turn under the grading rules. The usual objective in pruning has been the production of clear timber, but the FPRL study and other dissections of pruned logs have made it clear that under British rates of growth relatively little clear wood is likely to be obtained in the normal rotation by through-and-through sawing, though special tangential sawing, or rotary veneering, would certainly produce a higher proportion. What was striking however in the FPRL tests of logs pruned thirty or so seasons ago was the upgrading to the higher categories of the timber from the pruned logs. This, coupled with the removal of specific defects (such as black knots in pine) might be a more realistic motive for pruning than the production of clears.

Some indications of differences in timber properties due to provenance of seed have been obtained. In Lodgepole pine, coastal provenances appear as good in intrinsic qualities as continental provenances; there is still however the problem of poor stem form which lessens the useful yield. In Sitka spruce southerly provenances such as Oregon and Washington exhibit much faster growth than the standard Queen Charlotte Island material, but higher yields seem likely to be accompanied by some reduction in density, and a greater tendency to drought crack; however the total dry matter production is still higher in the southern provenances. As mentioned in Chapter 13, the most important criterion in the selection of "plus trees" in the spruce breeding programme is density, and some notable advances have been made at Princes Risborough in techniques for studying cores, which are proving useful in breeding work, and in enquiries on the effects of different silvicultural practices on wood development.

No attempt is made here to review the great amount of development work on home-grown timbers which has been undertaken by the Forest Products Research Laboratory in the post-war period. It has covered a wide range of topics; sawmilling practice; veneering and plywood; grading; seasoning; wood preservatives etc. Important work on timber of all sources has also been undertaken by the Timber Research and Development Association.

The Utilisation Development Section has maintained a close interest in all this work, and one of its major objects has been to ensure that wherever the results of research have shown that home-grown timber is suited to any particular purpose, it is not excluded explicitly or implicity in regulations or specifications.

REFERENCES

AARON, J. R. (1970). *The utilisation of bark*. Res. Dev. Pap. For. Commn, Lond. 32.

CROWTHER, R. E. (1964). *Extraction of conifer thinnings*. Bookl. For. Commn 11.

CROWTHER, R. E., and TOULMIN-ROTHE, I. (1963). *Felling and converting thinnings by hand*. Bookl. For. Commn 9.

FORESTRY COMMISSION (1956). *Report of the Committee on Marketing Woodland Produce 1956*. London: HMSO.

FORESTRY COMMISSION (1959). *Small pulp mill survey, Economic study, United Kingdom*. By Sandwell & Co. Ltd., Consulting Engineers, Vancouver, Canada .London: HMSO.

HARDING, T. (1969–1972). The joint programme on home-grown timber: Forest Products Research Laboratory and Forestry Commission. *Rep. Forest Res., Lond.* 1969, 134–138; 1970, 162–167; 1971, 118–123; 1972, 134–138.

HENMAN, D. W. (1963). *Pruning conifers for the production of quality timber*. Bull. For. Commn, Lond. 35.

HOLTAM, B. W. (1966). *Home-grown roundwood*. Forest Rec., Lond. 52.

MCNULTY, M. E. (1951). Nursery mechanisation. *J. For. Commn* 21 (1950), 52–53.

RICHARDS, E. G. (1967). *Appraisal of national wood production and consumption trends and their interplay with regional and world trends*. Res. Dev. Pap. For. Commn, Lond. 33.

RICHARDS, E. G., and TINSON, E. J. F. (1967). *Some problems of long-term marketing arrangements*. Res. Dev. Pap. For. Commn, Lond. 62.

ROSE, A. (1963). The Ledmore mounted lining-out plough Mark III. *J. For. Commn* 31 (1962), 90–97.

RYLE, G. B. (1969). *Forest service: The first 45 years of the Forestry Commission*. Newton Abbot, Devon: David & Charles.

Chapter 28

PUBLICATION AND RELATED SERVICES

The main traffic in written information has been the responsibility of two sections; Publications and the Library. H. L. Edlin has been in charge of publications since 1945, and for most of the period has answered to the Director of Research. The Library at Alice Holt (the Commission's library rather than a purely research facility) was based on a limited number of technical books and other documents transferred from Savile Row, and was rapidly expanded under G. D. Kitchingman. It is now in the charge of an officer with general responsibility for research information and liaison, O. N. Blatchford. The Photographic Section, which dates from 1949 and has been under I. A. Anderson for the rest of the period, has contributed equally to post-war publications and other modes of communication: lecturing, demonstration, exhibits, etc.

It is fashionable to speak of an "explosion" in scientific information during the last few decades, though a biological analogy with the rapidly ascending part of a growth curve would be a better image, and might offer some hope to librarians. Forest science has had its share in the problems connected with this rapid growth, probably due more to the widening of its sphere of interest than to the increased volume of work on strictly forest topics, though that itself is apparent enough.

The Library now contains some 6,000 books, besides many periodicals, separates and other documents. Besides its primary function as a reference and lending library, a great deal of documentation work has been undertaken, especially in general, regional and historic forestry topics. The basis for subject indexing has been from the outset the Oxford Decimal Classification (Commonwealth Agricultural Bureaux, 1954), an elaboration of the Universal Decimal Classification which was pioneered by the Swiss Flury, and further developed by F. C. Ford Robertson, Director of the Commonwealth Forestry Bureau, and his colleague (now the Director) P. G. Beak.

The choice of Alice Holt as the Commission's (first) research station has entailed some extra effort in library and documentation services which would not have been necessary had the station been sited at Oxford adjacent to the Institute's library, and also to the Commonwealth Forestry Bureau, which provides the principal forestry documentation service in the English speaking world (Ford Robertson, 1962). There has, however, been some gain in speed and selectivity in working directly from periodicals and reports etc.; and the Bureau's publication *Forestry abstracts* and card service has been of primary importance as *the* comprehensive scan of world literature in forestry and related subjects.

The number of periodicals received at Alice Holt has increased steadily over the period and now exceeds 300 titles. The majority of these are received under exchange arrangements with corresponding institutions, but the widening of forest research makes it necessary to take more pure scientific literature, and there are also a growing number of outright specialisations to be catered for.

The problem in a documentation service for a research establishment is where to stop. At Alice Holt some specialised fields (notably forest pathology and entomology) have been left to the Sections concerned, but a very large card index in the central forestry topics has been built up. In retrospect this documentation work now seems to have been over-ambitious. With limited resources it is impossible to be comprehensive, and it may be doubted whether wide coverage of individual items short of this is worth while. It is better to concentrate on providing the keys to documentation; i.e. references to the authoritative indexes, important bibliographies and critical reviews of literature. These aspects of documentation and information service are now being given increasing attention. The other aspect of this work is "current awareness"—drawing people's attention to the latest important publications, accession of books etc.

The importance of liaison work need hardly be stressed. Since the war several new organisations concerned with private forestry have come into being, and research with a bearing on forestry is now to be found in a variety of institutions—not merely the obvious ones. Liaison at the international level is now well catered for; the oldest institution, the International Union of Forest Research Organisations, has been very active in the period, and a number of Commission and university research people have served on its subject groups and working parties. IUFRO held its Congress in Britain in 1956, and James Macdonald had the honour of presiding at the subsequent Congress in Vienna in 1961. Other important international conferences have been held at regular intervals by FAO and the Commonwealth Forestry Association, and though general, these have usually featured research. Besides improving contacts and providing for exchanges of ideas, such conferences often stimulate publication, sometimes of matter which individual countries might not put together without such stimulus; the series of reports on experiences with exotic trees prepared for the 1957 Commonwealth Conference is one such example.

Dissemination of research information is a part of many activities "for the purpose of promoting forestry", which are too general to discuss here: advisory work; conferences; exhibitions; and lecturing to all types of audience. The contribution of the Photographic Section has been of great value in these activities, especially the large collection of high quality colour transparencies which covers nearly all aspects of home forestry, and has a fair representation also of forestry abroad.

The biggest load of advisory work has fallen on the pathologists and entomologists, who have to combine the roles of research workers with that of consulting specialists. Whilst this has undoubtedly lessened the time available for research, running an advisory service does recruit an army of observers, and the analysis of complaints often gives useful pictures of the distribution of troubles, especially those prevalent in particular types of weather.

Publication is the accepted formal means of disseminating research information, but till comparatively recently a good deal has been put out in limited circulation. Before the war, and for some time after it, there was a tendency to distinguish between completed work with "hard" results, and work in progress or which was regarded as merely indicative in deciding what to publish. James Macdonald's decision to publish the *Report on forest research* (the first issue covered the year ended March 1949) marked a change in view, and it has remained the principal means of reporting progress in the Commission's own and in grant-aided work since then. Previously, the *Journal of the Forestry Commission*, which was printed for internal circulation, carried a good deal of research information, but this declined after the publication of the *Report* and the *Journal* ceased issue in 1969. The Commission's two major pre-war series, Bulletins and Leaflets, have retained their original aims; the former being used to carry reports on major investigations, and the latter information on topics of technical or quite general interest. Both series continue to carry research or other material written by authors outside the Commission.

Besides the *Report on forest research*, two new series were introduced after the war, principally to carry research information. The Forest Records first appeared in 1949, and have usually reported individual investigations of a limited extent, though the series has carried straight technical information such as Yield and Volume Tables. It has been a full priced publication series from the outset. The other series, Research Branch Papers, latterly titled Research and Development Papers, started in 1951 as mimeographed notes of no great length for internal circulation, the object being to get miscellaneous research reports into a convenient format, replacing un-numbered, unlisted, mimeographed documents, which usually got lost. From such mundane beginnings, the R. and D. Papers became tidily printed and produced jobs, and latterly emerged as full publications. They are unpriced and used for matter which does not warrant wide circulation.

Between 1954 and 1964 the *Report* itself carried papers in addition to the normal accounts of progress, but this practice was dropped as there appeared to be adequate means of publishing such material elsewhere; also the *Report* was by this time becoming rather bulky.

A notable feature of the post-war publication has been the increase in the numbers of periodicals in which papers originating in forest research have been printed. The four important forestry journals published in Britain (*Scottish Forestry*, *Quarterly Journal of Forestry*, *Forestry*, and the *Commonwealth Forestry Review*) have always taken research matter, though scientific papers in the strict sense are usually confined to the two last mentioned. There has been a growing tendency however for specialists to publish in periodicals devoted to their own disciplines, of which there are now a large number, many being international in coverage and readership.

Broadly speaking, the choice of venue for publication seeks the best circulation for the article concerned; the interest may be general, regional, or specialised. However, there are great advantages in maintaining a good coverage of the central topics in the Commission's own series, so that they represent a convenient working library for the bookshelf; and this policy has been maintained. In research, the scientific paper of limited range has become the standard means of communication, and separates are scattered round like confetti. In lengthy and wide ranging investigations there is the problem of putting it all together, which may be an extremely onerous undertaking, and there is a tendency to rest on a series of short papers provided they meet scientific publication conventions, and are not merely progress reports. However, so much being contained in numerous papers there are problems of digestion, and one of the most valuable forms of literary endeavour is the critical review, which is well exemplified in the leaders in *Forestry abstracts*.

Review style papers for conferences also may be very successful modes of communication; they are perforce short, and not being scientific papers in themselves are not cumbered with detailed evidence. One type of paper which is not commonly seen deals with the interactions between different fields of research. An effort to write such a paper about forest research in the wide sense would probably show that the linkages between many lines are well appreciated, but it is probable that it would also reveal examples of compartmenting in which a line is being pursued without full recognition of the implications of other work. In any case, it seems evident that if the primary results of research' are going to be published in numerous short contributions, there will be an increasing need for communications which interpret and evaluate research against a broader background.

REFERENCES

COMMONWEALTH AGRICULTURAL BUREAUX (1954). *The Oxford system of decimal classification for forestry.* Being the definitive English version as authorised by the Rome Congress of the International Union of Forest Research Organizations, Sept. 1953, and published on their behalf by the Commonwealth Agricultural Bureaux, Farnham Royal, Bucks., England.

COMMONWEALTH FORESTRY ASSOCIATION (1962). *Commonwealth forestry handbook.* London. See also Empire Forestry Association, following.

COMMONWEALTH FORESTRY BUREAU, OXFORD (1950). *Guide to the use of Forestry Abstracts.* Commonwealth Agricultural Bureaux. Also second edition, 1958. (See also Imperial Forestry Bureau, following.)

EDLIN, H. L. (1966). *Check list of Forestry Commission publications, 1919–65.* For. Rec., Lond. 58.

EMPIRE FORESTRY ASSOCIATION (1930). *Empire forestry handbook.* Later editions in 1931, 1933, 1938, 1946, 1952 and 1957. Empire Forestry Association, London. A reference book covering, internationally, forestry departments, research institutions, and periodicals, plus a valuable glossary of commercial timbers based on British Standards 881 and 589: 1955. Continued as *Commonwealth forestry handbook,* 1962 (see above).

EMPIRE FORESTRY ASSOCIATION (1953; 1957). *British Commonwealth forest terminology.* Part 1 (1953): Silviculture, protection, mensuration and management, together with allied subjects. Part 2 (1957): Forest products research, extraction, utilization and trade. Empire Forestry Association, London.

FORD ROBERTSON, F. C. (1954). History of the international classification of forest literature. *Forestry Abstracts* **15** (2), 137–139. A review article.

FORD ROBERTSON, F. C. (1962). *The work of the Commonwealth Forestry Bureau, Oxford,* 1957–62: *a progress report.* Forestry Commission, London. (8th British Commonwealth Forestry Conference, East Africa).

FORD ROBERTSON, F. C. (Editor) (1971). *Terminology of forest science, technology, practice and products.* Washington, D.C. Society of American Foresters.

FORESTRY ABSTRACTS. See Imperial Forestry Bureau, following.

FORESTRY COMMISSION. *Catalogue of Publications.* 1973.

IMPERIAL FORESTRY BUREAU, OXFORD (1939–). *Forestry Abstracts,* 1939–to date. Currently edited by the Commonwealth Forestry Bureau, Oxford, and published by the Commonwealth Agricultural Bureaux, Farnham Royal, Bucks., England.

KITCHINGMAN, G. D. (1952). Basic list for Conservancy libraries. *Library Record, Forestry Commission Research Station, Alice Holt.* No. 13, 120–124.

KITCHINGMAN, G. D. (1954). The problem of finding the information you want. *Q. Jl For.* **48** (4), 277–280. A note on forestry documentation with particular reference to the older English literature and the means of making it accessible.

MAYNE, P. (1973).* *Check-list of Forestry Commission publications* 1919–1973. Forestry Commission Research and Development Paper No. 100.

OXFORD SYSTEM OF DECIMAL CLASSIFICATION FOR FORESTRY, THE. See: Commonwealth Agricultural Bureaux, above.

* A complete record of publications issued by the Forestry Commission itself, either directly or through HMSO. Does not include contributions to independent journals, etc. Copies available from Forestry Commission, 25 Savile Row, London W1X 2AY.

Chapter 29

CONCLUSIONS

Although some concluding remarks are called for, these can certainly not take the shape of a formal summary, since much of the material is over-compressed as it is. The practical results of research in various topics have also been the subject of comment, and any attempt at a general evaluation would be over-ambitious.

The growth in capacity and range of research during the period under review has been an obvious feature. Naturally, the greatest effort has been applied to problems of afforestation, and to the growth, yield and health of the plantations of the first rotation; and though most research (however practical its aims) is open-ended, the fifty-year period has rounded off many of the central enquiries concerned with these crops at a satisfactory level of knowledge. In the last two decades there has been a marked shift towards topics concerned with the older plantations.

If British forestry has good reason for satisfaction with the progress made in afforestation on many types of land—some of which have presented considerable difficulties—the future composition and structure of the forests still remains an open question. At present technological developments argue for a succession of even-aged pure crops in production forest, and so far no adverse evidence has been advanced in this country (though it has elsewhere). However, if for silvicultural reasons we do find it desirable to diversify our forests we are fortunate to have so many alternative species at our disposal, and it is a mistake to assume that technology can only be of use in the simplest silvicultures.

The recent emphasis on multi-purpose forest is not without interest in its influence on research. It will certainly widen the field, and some lines of enquiry may call for fresh approaches. Such topics as wildlife, recreation, amenity and integration with agriculture have already attracted attention, though the areas of enquiry have so far been limited, and not all questions about the use of woodland are primarily silvicultural, though the forester may be called on to answer them. In connection with agriculture, for instance, integration is usually considered in terms of shelter and the share-out of the land surface between plantations and grazing land. But obviously more intimate patterns are conceivable and might warrant enquiry should there be any theoretical advantages in them.

It is worth making the point that knowledge gained in forest research directed to productive woodland is often applicable to the establishment and management of woodland for other purposes.

The general conduct of state-aided research has been under discussion at the highest level in recent years. The Forestry Commission's own arrangements have been described; over the whole period the Commission's own research has been supported by grant-aided investigations at outside institutions, and only towards the end of the period has there appeared a Research Council with forest science in its terms of reference. Lately also there has been a movement towards separating developments from research, together with services which have evolved from enquiries started under the aegis of the Research Branch.

The most profitable relationships experienced by the Commission with other institutions might be described as alliances; a body such as the Commission carrying out its own objective research is well placed in this respect since, besides co-operating in the investigations, its staff has been able to interpret basic work at the practical level, and where necessary "scale-up" through field experimentation. This is much more than the "D" relationship to "R", for the better the objective research the more likely it is that profitable questions will be engendered for the pure specialist.

It has been common experience that the problem as it appears to those who are first confronted with it may take a very different shape once a new line of research has been started on it. This is of course saying no more than that the recognition of a difficulty does not necessarily pose the right research questions, but it does suggest that a good deal of flexibility should be allowed in terms of reference.

Besides questions of the conduct and control of state-aided research, on which policy decisions are likely to have been reached before this is printed, the future size of the forest enterprise must of course have an influence on forest research. This also is a matter of top-level policy discussion at the time of writing, and no comment on the various arguments will be expected here. It seems probable that forest policy, like agricultural policy, will become more "European" and less domestic. No doubt future research expenditure will relate to the size of the national investment in forestry, but it has become a flexible and pragmatic system, well adapted to changes in direction and scope.

ABBREVIATIONS USED, WITH ANNOTATIONS

ACLAND COMMITTEE. Ministry of Reconstruction: Reconstruction Committee: Forestry Sub-Committee. Presented Final Report to Parliament as Cd. 8881, 1918. *Chairman:* The Rt. Hon. F. D. Acland, M.P.

ADVISORY COMMITTEE ON FOREST RESEARCH. A Forestry Commission committee set up in 1930.

ARC. Agricultural Research Council. Established 1931.

BEDFORD COLLEGE. University of London (Regents Park, London N.W.1).

BRITISH EMPIRE FORESTRY CONFERENCES. 1920 (First) to 1947 (Fifth). Continued as British Commonwealth Forestry Conference 1952 (Sixth) to 1962 (Eighth). Continued further as Commonwealth Forestry Conferences, 1968 (Ninth) to date.

COMMONWEALTH FORESTRY CONFERENCES. *See* British Empire Forestry Conferences.

COMMONWEALTH FORESTRY INSTITUTE. *See* Imperial Forestry Institute.

DSIR. Department of Scientific and Industrial Research. Operated from 1915 to 1965. Responsible for Forest Products Research Laboratory.

FOREST PRODUCTS RESEARCH LABORATORY. *See* Princes Risborough Laboratory.

HAFOD FAWR. Near Penrhyndeudraeth, Merioneth. Now part of Beddgelert Forest. An early experimental plantation of the Office of Woods.

IMPERIAL FORESTRY INSTITUTE. Now Commonwealth Forestry Institute. University of Oxford (South Parks Road, Oxford).

IMPERIAL INSTITUTE OF BIOLOGICAL CONTROL. Now Commonwealth Institute of Biological Control, Gordon Street, Curepe, Trinidad, West Indies.

INSTITUTE OF TREE BIOLOGY. Established by the Natural Environment Research Council at the Bush Estate, Roslin, near Edinburgh, in 1972.

JEALOTT'S HILL. Jealott's Hill Research Station, Bracknell, Berkshire. (Imperial Chemical Industries, Ltd.)

KEW. Royal Botanic Gardens, Kew, Richmond, Surrey. (Ministry of Agriculture, Fisheries and Food.)

MACAULAY INSTITUTE. Macaulay Institute for Soil Research, Craigiebuckler, Aberdeen.

NATURE CONSERVANCY. Established 1950. Responsible to the Natural Environment Research Council: 1965–73 (Reconstituted 1973).

NERC. Natural Environment Research Council. Established 1965.

OFFICE OF WOODS. Office of Woods, Forests and Land Revenues, under His Majesty's Commissioners. Responsible for administration of Crown forests until 1924 (Transfer of Woods Act 1923).

PROFESSIONAL CLASSES OF THE WORKS GROUP OF CIVIL SERVANTS. Now the Senior Grades of the Professional and Technology Category of the Civil Service.

PRINCES RISBOROUGH (LABORATORY). Now part of the Building Research Establishment, Department of the Environment. Formerly responsible to the Department of Scientific and Industrial Research and later to the Ministry of Technology. (Princes Risborough, Buckinghamshire.)

RESEARCH ADVISORY COMMITTEE. *See* Advisory Committee on Forest Research.

ROTHAMSTED. Rothamsted Experimental Station, Harpenden, Hertfordshire (Lawes Agricultural Trust).

SOCIETY OF FORESTERS. Now the Institute of Foresters. (Newton House, Newton of Falkland, Freuchie, Fife.)

INDEX

Printed in Scotland for Her Majesty's Stationery Office by R. & R. Clark Ltd., Edinburgh

Dd 504045 k48 2/74